# Regulating Shale Gas

LEUVEN GLOBAL GOVERNANCE

**Series Editor:** Jan Wouters, *KU Leuven, Belgium*

The Leuven Global Governance series comprises high-quality monographs and edited volumes dedicated to innovative research on global governance. The series covers a wide range of themes, from the EU and global multilateral governance to trade, sustainable development, peace and security, human rights, democracy and rule of law, private standards, and global commons. In full recognition of the complex issues involved, the series favours studies approaching global governance from a multi-disciplinary, multi-level and multi-actor perspective.

Titles in the series include:

China, the European Union and Global Governance
*Edited by Jan Wouters, Tanguy de Wilde, Pierre Defraigne and Jean-Christophe Defraigne*

Private Standards and Global Governance
Economic, Legal and Political Perspectives
*Edited by Axel Marx, Miet Maertens, Johan Swinnen and Jan Wouters*

The Governance of Climate Relations between Europe and Asia
Evidence from China and Vietnam as Key Emerging Economies
*Edited by Hans Bruyninckx, Qi Ye, Nguyen Quang Thuan and David Belis*

Indirect Expropriation in International Law
*Sebastián López Escarcena*

China, the European Union and the Developing World
A Triangular Relationship
*Edited by Jan Wouters, Jean-Christophe Defraigne and Matthieu Burnay*

Global Governance and Democracy
A Multidisciplinary Analysis
*Edited by Jan Wouters, Antoon Braeckman, Matthias Lievens and Emilie Bécault*

The Contribution of International and Supranational Courts to the Rule of Law
*Edited by Geert De Baere and Jan Wouters*

Global Governance of Labor Rights
Assessing the Effectiveness of Transnational Public and Private Policy Initiatives
*Edited by Axel Marx, Jan Wouters, Glenn Rayp and Laura Beke*

Global Governance through Trade
EU Policies and Approaches
*Edited by Jan Wouters, Axel Marx, Dylan Geraets and Bregt Natens*

Regulating Shale Gas
The Challenge of Coherent Environmental and Energy Regulation
*Leonie Reins*

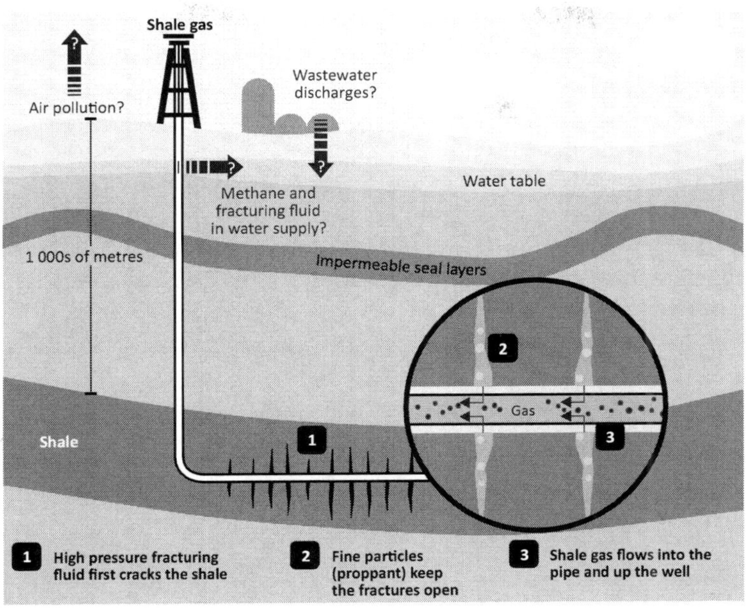


*Source:* International Energy Agency, *Golden Rules for a Golden Age of Gas – World Energy Outlook Special Report on Unconventional Gas* (2012), at 25.

*Figure 1.2 The shale gas exploration process and associated environmental impacts*

risks to the environment.[55] Extraction can adversely affect various aspects of the environment such as water resources, freshwater wetlands, eco-systems and wildlife, air quality, noise level, the seismicity of the rocks and can cause visual impacts on the site.[56] This introduction will not

55 See Bailey, note 53 above, at 819ff, also for further impacts on the environment.

56 For a detailed analysis of the factors, see Department of Environmental Conservation, New York State, 'Chapter 6, Potential environmental impacts' in: *Draft Supplemental Generic Environmental Impact Statement on the Oil, Gas and Solution Mining Regulatory Program – Well Permit Issuance for Horizontal Drilling and High-Volume Hydraulic Fracturing to Develop the Marcellus Shale and Other Low-Permeability Gas Reservoirs* (2009), as well as J. T. Bartis, T. LaTourrette, L. Dixon, et al., *Oil Shale Development in the United States: Prospects and Policy Issues* (Santa Monica, CA: RAND Corporation, 2005) at 35ff.

describe nor assess all possible adverse impacts and risks of the technology, but will provide three brief examples.[57]

First, regarding air pollution, even if the footprint of shale gas has not been at the centre of the scientific investigation yet, it has been argued that shale gas is not as 'clean' as has been assumed by some scholars and policy-makers, who referred to shale gas as a 'bridge technology' for a low-carbon economy.[58] According to some scientists,[59] shale gas extraction can even exacerbate global warming and has a larger footprint than the production of coal and conventional gas.[60] Emissions from shale gas development arise from drilling, completion and production activities.[61] During the fracturing and drilling process itself, around 3–7 per cent of the methane in shale gas is believed to escape into the air. The emissions mostly derive from flowback water and fluids. Regarding conventional gas resources, the escaping methane emissions are estimated to range between 1–6 per cent.[62] Not to be disregarded are also the emissions resulting from increased trucking infrastructure for the transport of clean and waste water, sand and equipment and materials.[63]

---

57 For a detailed description of the risks associated with shale gas see also T. J. Centner and N. S. Eberhart, 'The use of best management practices to respond to externalities from developing shale gas resources' *Journal of Environmental Planning and Management* 55(11) (2015), 4978–90 at 4980ff, as well as UBA, *Umweltauswirkungen von Fracking bei der Aufsuchung und Gewinnung von Erdgas insbesondere aus Schiefergaslagerstätten Teil 2 – Grundwassermonitoringkonzept, Frackingchemikalienkataster, Entsorgung von Flowback, Forschungsstand zur Emissions- und Klimabilanz, induzierte Seismizität, Naturhaushalt, Landschaftsbild und biologische Vielfalt*, (2014), available at www.umweltbundesamt.de/sites/default/files/medien/378/publikationen/texte_53_2014_umweltauswirkungen_von_fracking_0.pdf (last accessed 06.10.2015), at AP 125ff, and at APR 41ff.

58 M. R. Cady, 'Drilling into the issues: a critical analysis of urban drilling's legal, environmental and regulatory implications' *Texas Wesleyan Law Review* 16(1) (2010), 127–58, at 146f.

59 For example, see Council of Scientific Society Presidents, 'Letter from the council to President Obama and senior administration officials', 04.05.2010, available at www.eeb.cornell.edu/howarth/CCSP%20letter%20on%20energy%20&%20environment.pdf; last accessed 26.05.2015.

60 See Bartis et al., note 56 above, at 40.

61 For a detailed analysis and calculation of the emissions, see Department of Environmental Conservation, note 56 above, at 60f.

62 R. W. Howarth, R. Santoro and A. Ingraffea, 'Methane and the green house-gas footprint of natural gas from shale formations – A letter' *Climatic Change* 106(4) (2011), 1–12, at 1.

63 Bailey, note 53 above, at 819ff.

In addition, the hydraulic fracturing process can change the seismicity of the rock permanently, which can lead to seismic events or earthquakes and further impact (local) water resources. Depending on the individual rock formation, between three and four million gallons of fresh water per well can be required, which – depending on the area – can have significant impacts on the local water capacity and lead to an over-use of local water resources.[64] Concerns associated are the adverse impact of water withdrawals, waste water treatment and disposal, as well as impacts on the local drinking water supply. Furthermore, the use of water and chemicals during the drilling and fracturing phase can lead to water pollution through flowback and from drilling fluids. Additional concerns relate to the pollution of surface and groundwater due to surface spills and leaks in pits which store the water for disposal or recycling.[65] Depending on the location and surroundings of the shale, groundwater usage can reach up to 4590 per cent.[66] The fracturing process can result in large amounts of flowback water from the treatment itself, or in produced water, which is concentrated subterranean saltwater from the rock formation brought to the surface during the drilling process.[67] The flowback water that returns to the surface after the treatment has been mixed with fracturing fluids and can be contaminated with chemicals.[68]

Third and finally, the methods of the waste and drilling water treatment depend on the individual shale, but three options are available: underground injection; treatment and discharge; or recycling.[69] Underground injection, being the easiest and most often first choice, requires a porous and permeable injection zone with reserve and storage pits near the fracturing area. Concerns are the contamination of groundwater due to leakage of those pits or spills, as well as the destruction of flora and fauna on the surface.[70] In cases where such zones are not available, the produced water is transported to either municipal waste water treatment plants or commercial treatment facilities.[71] Accidents and spills due to

---

64 Ibid.

65 Cady, note 58 above, at 138f, as well as Department of Environmental Conservation, note 57 above, at 3f.

66 Ibid., at 139.

67 Ibid., 140f, as well Arthur et al., note 52 above, at 19.

68 For more information, see Cady, ibid., at 140f.

69 Bailey, note 53 above, at 819ff.

70 Cady, note 58 above, at 140f, as well as Arthur et al., note 52 above, at 19.

71 Arthur et al., note 52 above, at 21.

transport on highways and through residential areas are concerns associated with this method.[72] The third method, to recycle the water, is only an option for drilling water, as waste water cannot currently be recycled due to the high degree of corrosiveness and contamination. With regard to the amount of water needed for the operation, this method seems to be reasonable and appropriate, but is still limited due to inefficiency, the limited amount of recycling capacity and economic reasons, as the disposal of drilling water in pits is less expensive.[73]

## 3. METHODOLOGY

### 3.1 Research Questions and Overall Structure

This volume examines the possibility of a coherent regulation of energy and environment in the EU – using shale gas as an example. Accordingly, the overall research question is:

> To what extent does the regulation of 'new' technologies such as shale gas highlight the challenges of a coherent regulation of energy and environment in the European Union?

In order to answer the questions raised in the introductory chapter, as well as the overarching research question, this volume is divided into four chapters in total. Each chapter will answer one or more research questions and related sub-questions.

Chapter 2 addresses the legal basis for environmental and energy regulation at EU level. It will examine the horizontal and vertical division of competence relating to shale gas. It examines the impacts on a CEP which shale gas might bring about. This chapter answers the following questions and related sub-questions:

- What is the legal relationship between Title XX on the Environment and XXI TFEU on Energy in terms of energy security of supply concerns and environmental concerns: are they really mutually compatible or does one take precedence over the other?
  - o To what extent does the Union have competence to act regarding shale gas?
  - o How does this relationship play out on a Member State level

[72] Cady, note 58 above, at142f.
[73] Ibid., at 146f.

regarding each State's right to determine the conditions for exploiting its energy resources?

- How will the issue of shale gas affect the Union's quest for a Common European Energy Policy in the absence of a Treaty provision?
- Do the energy title and its opt-out mechanism in the second paragraph prevent a Common Energy Policy (on unconventional gas resources) in the future?

Chapter 3 deals with risk and regulatory design in the EU. It compares the regulation of other 'new' technologies such as CCS and nanotechnology with shale gas and establishes which lessons can be learned from shale gas. It will focus on the final two research questions:

- What are the lessons learned from regulatory design (failures) of technologies of the past?
- What is the 'new' in new technologies? And how does this matter for the regulation?

### 3.2 Selection, Structure and Criteria of the Comparative Cross-cutting Technologies

Chapter 3 of this volume compares the regulatory framework and history of shale gas technology with other cross-cutting or 'new' technologies, namely CCS technology and nanotechnology. These case studies have been selected on the grounds that each technology illustrates another problem and is subject to another form of regulatory design. The CCS technology serves as a point of reference for an integrated approach to regulation of a 'new' technology, while nanotechnology represents a clear application of a precautionary approach to regulation.[74] At EU level, all technologies have been regulated before the inclusion of a separate legal basis on energy, and hence have been subject to the old set-up of competences.

In addition, the technologies have been referred to as (and to some extent still are) 'new' technologies, although the regulatory approaches to them differ. Chapter 3 accordingly analyses the correlation of 'newness', 'bindingness of the measure', 'underlying regulatory principle (prevention/precaution)' and 'public debate', as well as the 'business case'.

---

[74] For a more detailed reasoning of the selection of the technologies, please refer to Chapter 3.1.

The chapter aims not at outlining the regulatory approaches taken towards nanotechnology and CCS in detail, but rather focuses on aspects which are relevant regarding the regulation of shale gas. Taking into account the overall aim of the chapter, namely to analyse the regulatory trend taken towards the regulation of 'new' technologies and to establish if there is a link between the regulatory design and newness, the following criteria have been established:

- Process and general direction of regulation
- Risk management and the precautionary and preventive principles
- Addressing the public debate
- The business case.

## 3.3 Demarcation of Scope

The volume is focused on the regulatory framework and related questions for shale gas and includes major developments in legislation and policy up until 1 May 2015.

# 2. In search of the legal basis for energy and environmental regulation at the EU level: the case of unconventional gas extraction and its impacts on a common European energy policy

## 1. INTRODUCTION

To adapt to newly emerging environmental and energy challenges, European environmental law has undergone many revisions and changes in the past 30 years.[1] Whereas European environmental law reached a strong standing in the European *acquis* a long time ago, the standing of 'European energy law' as such is not as cemented. This is mostly due to sovereignty issues of the Member States,[2] as well as different Member States focusing on their national policies in this field. The existence of an 'EU energy law' is questioned as such or, at least, the question is raised whether 'energy law' is not in fact general EU law with a special colouring as applied to the energy sector.[3] The different standing of these interlinked areas within the European *acquis* is but one reason why the successful regulation of all environmental aspects of human activities and the interaction between environmental and energy priorities in particular,

---

[1] See for example C. Hey, 'EU environmental policies: a short history of the policy strategies' in: *EU Environmental Policy Handbook* (European Environmental Bureau, 2005), 17–31; and also L. Hancher, 'Energy and the environment: striking a balance?' *Common Market Law Review* 26(3) (1989), 475–512; as well as K. Talus, *EU Energy Law and Policy: A Critical Account* (Oxford: Oxford University Press, 2013), at 21.

[2] R. Leal-Arcas and A. Filis, 'Conceptualizing energy security through an EU constitutional perspective' *Fordham International Law Journal* 36(5) (2013), 1225–301, at 1226.

[3] Talus, note 1 above, at 1.

has proved to be challenging.[4] The inclusion of the Title XXI on Energy in the TFEU changed the Union's competences and the legal basis upon which it can act in this regard. It still remains to be seen whether this new constitutional set-up is better suited to reconcile energy supply security and environmental protection concerns.[5] This volume provides further insights into this dilemma. One emerging subject in both areas of law which provides testimony to this challenge, is the extraction of unconventional gas resources.[6]

The political crisis in Ukraine has raised questions about the Union's reliance on gas imports from Russia. The large-scale and manipulative nature of hydraulic fracturing techniques elicits a mixed response due to a large realm of unknowns, not only about externalities, but also concerning the ability of the technology to deliver.[7] Characterized by uncertainties and promises to address climate change, shale gas extraction also raises legal issues.[8] Shale gas exploration has two different angles inherent to it: one of energy exploration and supply security; and

---

[4] Especially regarding the areas of climate, air, land, biodiversity, waste or water. See European Commission, 'Energy and environment', available at http://ec.europa.eu/environment/integration/energy/index_en.htm; last accessed 28.06.2015.

[5] D. Benson and A. Jordan, 'A grand bargain or an incomplete contract?: European Union environmental policy after the Lisbon Treaty' *European Energy and Environmental Law Review* 17(5) (2008), 280–90, at 280.

[6] See also Talus, note 1 above, at 1.

[7] See for example differences in the estimation of the amount of resources in Poland: J. Teusch, *Shale Gas and the EU Internal Gas Market: Beyond the Hype and Hysteria* (CEPS Working Documents No. 369, 2012), at 3. The reference values can be expressed in 'in place' resources or technically recoverable resources, which can differ considerably in estimations and calculations. For an explanation of the two calculation methods, see also I. J. Andrews, *The Carboniferous Bowland Shale Gas Study: Geology and Resource Estimation* (London: British Geological Survey for DECC, 2013), at 5f and 10f, as well as DECC, 'Resources vs Reserves – What do estimates of shale gas mean?' 2013, available at www.gov.uk/government/uploads/system/uploads/attachment_data/file/256358/Publication_Resources_vs_Reserves_June13.pdf; last accessed 26.06.2015. See also C. McGlade, J. Speirs and S. Sorrell, 'Methods of estimating shale gas resources – Comparison, evaluation and implications' *Energy* 54(59) (2013), 116–25.

[8] See for example D. Forster and J. Perks, *Climate Impact of Potential Shale Gas Production in the EU* (AEA Consulting, 2012), as well as M. Ballesteros, F. Pelsy and L. Reins, *Regulatory Provisions Governing Key Aspects of Unconventional Gas Extraction in Selected Member States* (Milieu Ltd., 2013).

one of environmental concerns.[9] Within the EU, these twin aspects pose particular challenges. They put the synergy, or the lack thereof, between two distinct titles of the relevant Treaty to test.

This chapter focuses on the constitutional set-up at the basis of this case study. The inclusion of the Energy Title XXI in the Lisbon Treaty changed the nature of the EU's competences in the field of environmental and energy policy. The context of 'shale gas' constitutes a practical example of the issue playing out at two levels. First, the interrelation of the titles themselves and the assumption that the security of energy supply overrules environmental protection standards; and secondly, the individual Member States' policies and divergent approaches towards applying regulations on shale gas. In other words, the EU has to regulate a technology which clearly has cross-title impact, on the basis of two titles in the Treaty with competing objectives.[10]

This chapter answers the following research questions and related sub-questions:

- What is the legal relationship between Title XX on the Environment and XXI TFEU on Energy in terms of energy security of supply concerns and environmental concerns: are they mutually compatible or does one take precedence?
    - To what extent does the Union have competence to act regarding shale gas?
    - How does this relation play out at a Member State level regarding each State's right to determine the conditions for exploiting its energy resources?

---

[9] See for the challenge of combining these aspects in general (not related to shale gas): S. Gaines, 'The energy revolution as sustainable development' in: L. Squintani and H.H.B. Vedder (eds), *Sustainable Energy United in Diversity: Challenges and Approaches in Energy Transition in the EU* (European Environmental Law Forum, 2014), 7–24.

[10] This chapter is partly based on L. Reins, 'Shale gas in the European Union: You, me, together? Reflections from a subsidiarity perspective' *Oil, Gas, Energy Law Intelligence* 12(3) (2014), 1–11, L. Reins, 'In search of the legal basis for environmental and energy regulation at the EU level: the case of unconventional gas extraction' *Review of European Community & International Environmental Law* 23(1) (2014), 125–33, as well as L. Reins, 'European minimum principles for shale gas: preliminary insights with reference to the precautionary principle' *Environmental Liability* 22(1) (2014), 16–27.

- How will the issue of shale gas affect the Union's quest for a Common European Energy Policy in the absence of a Treaty provision?
  - o Do the energy title and its opt-out mechanism in the second paragraph prevent a Common Energy Policy (on unconventional gas resources) in the future?

This chapter first provides an introduction to the changes brought about by the Lisbon Treaty to the EU's energy and environmental competences, as well as to supply security. Then, the vertical and horizontal set-ups of competences with regard to shale gas are analysed. Lastly, the chapter will focus on the implications of shale gas on a Common European Energy Policy.

## 2. 'ENERGY' AND 'ENVIRONMENT' BEFORE AND AFTER THE LISBON TREATY

This section first provides an overview of the nature and interaction of the two policy areas prior to the introduction of the Lisbon Treaty, focusing on the their legal bases. It will not discuss the history and development in its entirety, as this has previously been done elsewhere.[11]

### 2.1 The Environmental Legal Basis

In 1986, a separate legal basis for the environment was introduced into the Single European Act (Articles 130r–130t). This does not mean that environmental action was not dealt with at a European level prior to this. Before the introduction of a separate legal basis on the environment, environmental issues were to a large extent dealt with under the headings of Articles 100a and 235 EEC, respectively (Common Market (later: 'Internal Market', now Article 114 TFEU)), and 'implied powers' (now Article 352 TFEU, also referred to as the 'flexibility clause').

---

[11] Talus, note 1 above, at 15–65 and 66–106; as well as G. Van Calster and L. Reins, *EU Environmental Law* (Cheltenham: Edward Elgar Publishing, forthcoming 2016).

One important difference from the introduction of the specific legal basis on the environment between the common market provision and the environmental one was the decision-making procedure. Whereas unanimity was needed for the environmental competence, majority voting was applicable for the common market. The Amsterdam Treaty in 1997 introduced co-decision-making regarding environmental issues, as well as the principle of a high level of environmental protection (Article 2 ECT) as a Union objective.[12] The Treaty further promoted the first 'environmental only' integration principle as a general principle of Union law. Hence it can be stated that 'by default', all EU policy is to a certain extent also environmental policy.[13] Thus, despite the existence of a special environmental legal basis, not all environmental policy is based upon it. Depending on the exact content of a measure, a wide variety of legal bases can be used: Article 43 TFEU on a common agricultural policy,[14] Article 100 TFEU on transport,[15] Article 113 TFEU on taxes,[16] Article 114 TFEU on the internal market,[17] Article 153 on social policy,[18] Article 182 TFEU on research and technological development,[19] Article

---

[12] See for a detailed discussion G. Van Calster and K. Deketelaere, 'Amsterdam, the IGC and greening the EU Treaty' *European Environmental Law Review* 7(1) (1998), 12–25.

[13] Van Calster and Reins, note 11 above, as well as N. Dhondt, *Integration of Environmental Protection into Other EC Policies: Legal Theory and Practice* (Groningen: European Commission Law Publishing, 2003), for a more detailed discussion.

[14] For example, Directive 91/414/EEC of 15 July 1991 concerning the placement of plant protection products on the market, [1991] OJ L 230/1.

[15] For example, Directive 2002/30/EC of 26 March 2002 on the establishment of rules and procedures with regard to the introduction of noise-related operating restrictions at Community airports, [2002] OJ L 85/40.

[16] For example, Council Directive 2003/96/EC of 27 October 2003 restructuring the Community framework for the taxation of energy products and electricity, [2003] OJ L 283/51.

[17] For example, Regulation (EU) No. 528/2012 of 22 May 2012 concerning the making available on the market and use of biocidal products, [2012] OJ L 167/1.

[18] For example, Directive 89/391/EEC of 12 June 1989 on the introduction of measures to encourage improvements in the safety and health of workers at work, [1989] OJ L 183/1.

[19] For example, for the adoption of the multiannual framework programmes ('FPs') promoting research also in the environmental area.

207 TFEU on common commercial policy[20] and Articles 196, 214 and 222 TFEU[21] on civil protection, humanitarian aid and solidarity and can be used for passing legislation addressing environmental issues.[22]

## 2.2 Absence of an Energy Chapter in the Treaties

The beginning of a 'European' energy policy was characterized by the inclusion of common objectives into national energy policies[23] in order to create a common understanding. Energy has always been a highly politicized policy area.[24] While having enjoyed special standing in European (legal) history,[25] energy has also been the force driving the establishment of the founding Treaties of what is now the European Union. At the time, the 1951 Paris Treaty establishing the European Coal and Steel Community (ECSC), and the 1957 EURATOM Treaty were considered to be a sufficient foundation for energy policy.[26] As a result of the energy crisis in the 1970s, the Member States were reluctant to delegate powers to the European Institutions, and consequently give up their 'energy sovereignty'. The Commission effortlessly lobbied for a separate Directorate General (DG) on energy and an energy chapter in the Treaties.[27] Subsequently, one of the first official documents recognizing the need to integrate environmental considerations into energy policy and promote rational and economic use of energy resources was introduced during this period. The 1975 Council Resolution on Energy and Environment[28] announced policy proposals on thermal discharges,

---

[20] For example, Directive 2001/37/EC of 5 June 2001 on the approximation of the laws, regulations and administrative provisions of the Member States concerning the manufacture, presentation and sale of tobacco, [2001] OJ L 194/26.

[21] To date no action has been taken under these competences.

[22] See more detailed Van Calster and Reins, note 11 above.

[23] Hancher, note 1 above, at 475.

[24] Leal-Arcas and Filis, note 2 above, at 1126.

[25] Talus, note 1 above, at 15.

[26] L. Hancher and F. M. Salerno, 'Energy policy after Lisbon' in: A. Biondi, P. Eeckhout and S. Ripley (eds), *European Union Law after Lisbon* (Oxford: Oxford University Press, 2012), 367–402, at 368.

[27] Talus, note 1 above, at 17, as well as T. Maltby, 'European Union energy policy integration: a case of European Commission policy entrepreneurship and increasing supranationalism' *Energy Policy* 55 (2013), 435–44, at 436f.

[28] Council Resolution of 3 March 1975 on energy and environment, [1975] OJ C 168/2.

sulphur dioxide and nitrogen oxides in order to show the linkage of these two policy areas and to implement policy regarding such in practice.

Nevertheless, Member States' general perceptions on energy policy did not change until the 1980s.[29] Governments realized that energy can be traded as a private good in competitive markets instead.[30] In the late 1980s the Commission based its energy strategy on Commission Directives, actions for failure to fulfil an obligation of a Directive and harmonization legislation[31] such as the Price Transparency Directive, the Electricity and Gas Transit Directive as well as the Hydrocarbon Directive. This yielded only mixed and tentative results.[32] Later, in the style of the US's 'common heritage' and 'open access' to oil pipelines and gas, the Commission introduced a similar concept in order to complete the internal market and strengthen the international competitiveness of the EU oil and gas players.[33] The institution thereby applied a cautious approach, allowing the Member States to gradually shift away from State monopolies towards privatization and liberalization of their energy systems.[34] The first effort by the Commission to introduce a separate legal basis for energy policy into the Treaty came during the negotiations of the Maastricht Treaty as an attempt to introduce the concept of supply security in the Community, and to complete the common market for energy.[35] The inclusion however failed, and the Commission decided to apply a dual approach in further promoting energy issues without a separate legal basis. First the Commission would seek the creation and adoption of sector-specific legislation relating to energy issues (oil/gas/electricity). If such legislation failed to progress quickly enough, the Commission would act under the ambit of competition law and would also take recourse to infringement proceedings before the European Court of Justice.[36]

---

[29] M. Kanellakis, G. Martinopoulos, T. Zachariadis, 'European energy policy – A review', 62 *Energy Policy* (2013), 1020–30, at 1020.

[30] Ibid., at 1024.

[31] Talus, note 1 above, at 22.

[32] Ibid., at 39.

[33] D. Helm, 'The European framework for energy and climate policies' *Energy Policy* 64 (2014), 29–35, at 30f.

[34] Ibid.

[35] European Commission, 'Regierungskonferenzen: Beiträge der Kommission' (Bulletin der Europäischen Gemeinschaften, Beilage 2/91, 1991), at 151f.

[36] Talus, note 1 above, at 65.

The next attempt to introduce a specific legal basis on energy issues was put forward in the negotiations on the Constitutional Treaty.[37] Considerable discussion in the Intergovernmental Conference (IGC) in October 2004 surrounded the inclusion and exact drafting of an energy provision. In the end, the provision was carefully drafted in order to guarantee the sovereignty of the Member States over their natural resources and over their policies relating to taxation.[38] The Constitutional Treaty included a special legal basis for energy in Article III-256 (section 10). The Lisbon Treaty copied the wording of the Article and introduced minor changes; for example the inclusion of solidarity considerations, which were absent in the Constitutional Treaty. Whether this inclusion is really only a minor and cosmetic change will be further discussed in sections 3.2.1 and 5.4 below.

## 2.3 Legal Bases for Energy Issues Prior to the Treaty of Lisbon

Energy is a sensitive policy area which touches the essence of a Member State's sovereignty.[39] In the absence of a separate Energy Chapter before the Treaty of Lisbon in 2009, energy issues have been dealt with under a variety of competences in the EC Treaty or previous Treaties. The most prominent Treaty base on energy issues was the internal market with its four freedoms.[40] The provision was included in the Single European Act (SEA) in 1986 under Article 8a (Article 14 TEC and now Article 26 TFEU). Article 95 TEC (now Article 114 TFEU) created the legal basis for legislation related to the objectives of the internal market (now Article 26 TFEU). The Commission used these internal market provisions as a legal basis for energy legislation, allowing it to be adopted

[37] The Treaty establishing a Constitution for Europe (TCE), was an attempt to create a consolidated constitution for the EU. It was signed on 29 October 2004, but never ratified. The Lisbon Treaty, after a period of reflection and with many changes, was created to replace the Constitutional Treaty. For more information, please refer to the European Commission's, 'A Constitution for Europe', available at http://europa.eu/scadplus/constitution/introduction_en.htm, last accessed 28.06.2015.

[38] See for a more detailed discussion: Hancher and Salerno, note 26 above, at 372.

[39] See also section 5.

[40] In Article 26(2) TFEU the internal market is defined as 'an area without internal frontiers in which the free movement of goods, persons, services and capital is ensured in accordance with the provisions of the Treaties'.

under qualified majority voting instead of unanimity.[41] The entire 'Third Package'[42] legislation was passed under the internal market competence.[43] Other legal bases used for energy measures were Articles 154–156 TEC (now Articles 170–172 TFEU) for trans-European networks (on energy policy guidelines and finance schemes rather than legal acts),[44] Article 81–88 TEC on competition, as well the supplementary competence clause contained in Article 308 TEC (now Article 352 TFEU).[45] Additionally, Article 175 TEC (now Article 192 TFEU) continues to serve as a basis for environmental matters.[46] Several pieces

---

41 E. D. Cross, B. Delvaux, L. Hancher, et al., 'EU energy law' in: M. M. Rogenkamp, C. Redgwell, et al. (eds), *Energy Law in Europe: National, EU and International Regulation (*Oxford: Oxford University Press, 2007), 225–392, at 228.

42 The package consists of two Directives and three Regulations aimed at opening the gas and electricity market in the EU: Directive 2009/72/EC concerning common rules for the internal market in electricity and repealing Directive 2003/54/EC, [2009] OJ L 211/55; Directive 2009/73/EC concerning common rules for the internal market in natural gas and repealing Directive 2003/55/EC, [2009] OJ L 211/94; Regulation (EC) No. 714/2009 on conditions for access to the network for cross-border exchanges in electricity and repealing Regulation (EC) No. 1228/2003, [2009] OJ L 211/15; Regulation (EC) No. 715/2009 on conditions for access to the natural gas transmission networks and repealing Regulation (EC) No. 1775/2005, [2009] OJ L 211/36; Regulation (EC) No. 713/2009 of the European Parliament and of the Council of 13 July 2009 establishing an Agency for the Cooperation of Energy Regulators, [2009] OJ L 211/1.

43 For a detailed discussion of the Third Package and how it impacts the European competences in the field of energy, please refer to Hancher and Salerno, note 26 above, at 375ff.

44 J. Kuhlmann, 'Kompetenzrechtliche Neuerungen im europäischen Energierecht nach dem Vertrag von Lissabon' 79 *EI Working Papers* (WU Vienna: Europainstitut, 2008), at 10.

45 See also J. C. Pielow and B. J. Lewendel, 'The EU energy policy after the Lisbon Treaty' in: A. Dorsman, et al., *Financial Aspects in Energy: A European Perspective* (Berlin, Heidelberg: Springer, 2011), 147–65, at 151, as well as Client Earth, *The Impact of the Lisbon Treaty on Climate and Energy Policy – An Environmental Perspective* (2010), at 8 and for a visual overview S. Fischer, 'Energie- und Klimapolitik im Vertrag von Lissabon: Legitimationserweiterung für wachsende Herausforderungen' *Integration* 1(10) (2009), 50–62, at 53.

46 See also Kanellakis, Martinopoulos and Zachariadis, note 29 above.

of legislation on climate change[47] and renewable energy[48] have been passed under this competence.[49]

The main objective of the introduction of a specific energy competence in the European Treaties was to ensure a clear division of competences between the Union and the individual Member States,[50] and to create a 'special cooperation modus operandi in greater transparency with energy matters' between the Union and its Member States.[51] In this sense, Article 194 TFEU lays down the competences through which the EU can adopt legal measures, but also sets its boundaries, through the second paragraph in particular.[52] However, even with a specific legal basis for energy, not all related issues are adopted under this competence. The same legal bases used for the adoption of energy issues pre-Lisbon are still used today. The competition provisions (Article 101–106 TFEU) and environmental protection (Article 192 TFEU) are of special importance, but also the legal bases on free movement. Public monopolies, undertakings entrusted with services in the general economic interest, and state aid are still used.[53]

---

[47] For example, the legislation relating to the European ETS: Directive 2003/87/EC of the European Parliament and of the Council of 13 October 2003 establishing a scheme for greenhouse gas emission allowance trading within the Community and amending Council Directive 96/61/EC, [2003] OJ L 275/32; or the legislation establishing the regulatory framework on CCS in the Union: Directive 2009/31/EC of 23 April 2009 on the geological storage of carbon dioxide and amending Council Directive 85/337/EEC, [2009] OJ L 140/114, and Regulation (EC) No. 1013/2006 on shipments of waste, [2009] OJ L 140/114.

[48] For instance Directive 2009/28/EC of 23 April 2009 on the promotion of the use of energy from renewable sources and amending and subsequently repealing Directives 2001/77/EC and 2003/30/EC, [2009] OJ L 140/16.

[49] This remains the case, even after Lisbon. The Treaty added a specific reference to climate change in Article 192 TFEU, as further discussed below.

[50] P. Bogdanowicz, 'Division of competences between the European Union and the Member States in the area of the European Union energy law' in: S. Doumbé-Billé (ed.), *Défis énergétiques et droit international* (Brussels: Larcier, 2011), 201–12, at 202.

[51] B. Nowak, *Energy Policy of the European Union: Chosen Legal and Political Aspects and their Implications for Poland* (Warsaw: WAiP, 2009) at 53.

[52] See also section 3.2.1.

[53] Talus, note 1 above, at 8f.

## 2.4 'Energy' and 'Environment' Post-Lisbon: Scope and Decision-making

The introduction of Title XXI on Energy in the TFEU changed the competences of the European Institutions, allowing them to act in both the environmental and energy fields. While the express legal provisions in the environmental domain did not change aside from the new reference to climate change in the objectives, the changes in the general provisions of the Lisbon Treaty will also affect environmental law-making.[54] The new reference to climate change in the environmental provisions strengthens the EU's political commitment with a view to taking a leadership role in addressing this issue of global importance.[55]

The majority voting in the Council[56] simplifies the process of achieving a majority, and thus the actual passing of legislation. This has positive effects for the decision-making process itself as well as for the amount of legislation passed on the basis of the environmental competence.[57] For certain sensitive issues listed in Article 192(2) TFEU, such as fiscal provisions, town and country planning and measures significantly affecting a Member State's choices between different energy sources and the general structure of its energy supply, the special legislative procedure still applies; this means that unanimity in the Council is required, unless the Council unanimously renders these issues subject to the ordinary legislative procedure.

The energy title reshuffles the competences relating to energy issues. The decision-making procedure is the ordinary legislative procedure (further discussed under the vertical competences). Because measures affecting energy policy were previously adopted based on either the environmental or the internal market competence, scholars tend to conclude that there are great overlaps between energy and climate

---

[54] M. Lee, 'The environmental implications of the Lisbon Treaty', *Environmental Law Review* 10(2) (2008), 131–8, at 131.

[55] Ibid.

[56] The Treaty of Lisbon introduced a voting system based on the principle of double majority of States and population. Accordingly, two types of majority are needed for a proposal to pass; first, a majority of countries (at least 15); and, second, a majority of the total EU population (countries in favour must represent at least 65 per cent of the European population). The new voting system has been applied since 31 October 2014.

[57] See Lee, note 54 above.

policy,[58] and that the relationship between environmental and energy policies and legislation is of an ambivalent nature.[59] With the introduction of Article 194 TFEU however, energy is now a separate and competing title aimed at ensuring '(a) the functioning of the energy market, [and (b) the] security of energy supply, [as well as (c) the promotion of] energy efficiency, energy saving and the development of new and renewable forms of energy; [and (d)] the interconnection of energy networks'. The reference to environmental considerations (energy policy shall have '*regard [to] the need to preserve and improve the environment*') not only takes into account the requirements of the integration principle included in Article 11 TFEU,[60] but also the obligation that energy policy must work towards the achievement of environmental objectives, and needs to preserve and improve the environment, instead of merely 'taking environmental issues into account'.[61] The energy sovereignty of the Member States in determining their energy mix is thus constrained by environmental measures (Article 194(2) 2nd sub-paragraph).[62] EREF's action was deemed inadmissible on 23 November 2015.[63]

A similar conclusion can be drawn with regard to the internal market. The separate energy title gives energy issues stronger standing within the European system. It shows that energy policy and supply security is not just an area of national concern any more, but is rather a matter of European concern with cross-border impact.[64]

---

[58] See W. Frenz and A. M. Kane, 'Die neue europäische Energiepolitik' *Natur und Recht* 32(7) (2010), 464–76, at 469.

[59] Ibid.

[60] Kuhlmann however, is of the opinion that the reference to the environment restates the integration principle: Kuhlmann, note 44 above, at 18.

[61] Client Earth, note 45 above, at 8.

[62] ECJ, Case T-694/14, *EREF v Commission*, case in progress.

[63] EREF, 'Case T-694/14 European Renewable Energies Federation vs. European Commission – action for annulment of Chapter 3.3.2 of the Guidelines for Environmental and Energy Aid 2014-2020', Press Release, available at www.eref-europe.org/wp-content/uploads/EREF-Press-Release-06-02-2015.pdf; last accessed 26.06.2015.

[64] Pielow and Lewendel, note 45 above, at 148.

The uncertainty regarding the nature of the competence of Article 194 TFEU is a reason why its practical effect still remains largely unclear. As stated in the introduction to this chapter, the case study of shale gas is one of the first examples where the new Treaty set-up with an interaction between the environmental and energy competences plays out in practice.

## 3. THE VERTICAL DIVISION OF COMPETENCES AND SHALE GAS

Before examining the horizontal division of competences for shale gas in greater detail, the division of competences between the Member States and the Union will be evaluated. It needs to be established whether the Union can act at all, and if so, whether it can '*put forward legislative proposals with legally-binding provisions*'. The positions of the Member States on shale gas legislation and policy are quite divergent (see further section 5.3 below), ranging from the banning of the activity in France, to exploratory drilling in Poland and the UK A common European position appears difficult to establish, as demonstrated by the Council's refusal to include a mandatory Environmental Impact Assessment (EIA) on shale gas in a revised EIA Directive.[65] The Commission has nonetheless established some minimum principles on shale gas in the form of a non-binding Recommendation.[66] At first sight, this is quite surprising. Government officials, NGOs and the Committee of the Regions (CoR) had all asked for a stringent regime due to the associated negative environmental impacts of the activity, the uncertainties of how and if these activities will evolve in practice and the large public opposition in the Member States.[67] However, if one steps away from the political dimension of the debate and concentrates on the legal perspective, it becomes immediately apparent that the hands of the European

[65] Procedural file 2012/0297 (COD), Amendment of Directive 2011/92/EU on the assessment of the effects of certain public and private projects on the environment.

[66] Commission Recommendation of 22 January 2014 on minimum principles for the exploration and production of hydrocarbons (such as shale gas) using high-volume hydraulic fracturing, OJ [2014] L39/72 ('Recommendation').

[67] See for example Committee of the Regions, Opinion of the Committee of the Regions on 'Local and regional authorities perspective on shale/tight gas and oil (unconventional hydrocarbons)', [2013] OJ C 356/23.

Institutions might be tied. Shale gas does not lend itself for a comparison with other environmental issues which easily pass the subsidiarity hurdle.

This section focuses on the legal aspects of the challenge outlined above. Taking into account that the energy and environmental competences of the Union are shared (Article 4 TFEU), several questions arise: does the Union have competence to introduce a legal measure on shale gas? If the Union has competence, how far does it reach from a subsidiarity and proportionality perspective? Is the recently published Commission Recommendation on minimum principles for unconventional hydrocarbons an escape route in this regard? Through the blocking of a mandatory EIA, the 'proactive shale gas Member States' have made it clear that European involvement in the issue is not welcomed. The underlying question is thus: to what extent can the EU engage the issue at all? Taking into account the foregoing, one has to ask two essential questions. First, why has the Commission acted at all? Second, why has it done so by means of soft law?

### 3.1 Why Did the Commission Act?

In order to determine why the Commission acted on shale gas, one has to go back to some basic principles of European law determining the competences between the Union and its Member States. It is part of the Union's very nature that there is close cooperation between the Member States, the Union and its institutions.[68] This can be referred to as cooperative federalism in which 'national and state governments work together in the same areas, sharing functions and therefore power'.[69] Thus, a key characteristic for this form of federalism is the principle of attributed competences. In the Union the principle of attributed powers or 'conferral' (Articles 4(1) and 5(2) TEU) is the fundamental principle in determining the division of competences between the Union

---

[68] See also J. Bast and A. von Bogdandy, 'The European Union's vertical order of competences: the current law and proposals for its reform' 39 *Common Market Law Review* 2 (2002), 227–68, at 236.

[69] M. D. Reagan, *The New Federalism* (Oxford: Oxford University Press, 1972), at 21, as cited in R. Schutze, 'Cooperative federalism constitutionalised: the emergence of complementary competences in the EC legal order' *European Law Review* 31(2) (2006), 167–84, at 168. See contrary to this perception G. De Búrca, 'Limiting EU powers' *European Constitutional Law Review* 1(1) (2005), 92–8, at 95 who is of the opinion that 'the [European] system cannot easily be described either as one of dual or co-operative federalism' as the Treaty provisions contain aspects of both.

and the Member States. In addition, once the competence is conferred upon the Union, the principles of subsidiarity (Article 5(3) TEU) and proportionality (Article 5(4) TEU) determine the extent thereof.[70]

The principle of conferral establishes whether a European Union competence *exists* at all: 'competences not conferred upon the Union in the Treaties remain with the Member States' (Article 4(1) TEU), and 'the Union shall act only within the limits of the competences conferred upon it by the Member States in the Treaties to attain the objectives set out therein' (Article 5(2) TEU).[71] The principle confirms that the Union does not have a general legislative competence, and that '[t]he scope of and arrangements for exercising the Union's competences shall be determined by the provisions of the Treaties relating to each area' (Article 2(6) TFEU). The principles of subsidiarity and proportionality determine the *extent* to which the Union can exercise the competence (Article 5(1) TEU). The subsidiarity principle[72] is included in Article 5(3) TEU and reads:

> Under the principle of subsidiarity, in areas which do not fall within its exclusive competence, the Union shall act only if and in so far as the objectives of the proposed action cannot be sufficiently achieved by the Member States, either at a central level or at the regional and local levels, but can rather, by reason of the scale or effects of the proposed action, be better achieved at Union level.

The limitation to the principle requiring proportionality is included in the next paragraph of the provision (Article 5(4) TFEU), and states that 'under the principle of proportionality, the content and form of Union action shall not exceed what is necessary to achieve the objectives of the Treaties'.

### 3.1.1 Shared competences

In EU law, there are generally three categories of competence: exclusive (Article 3 TFEU), shared (Article 4 TFEU) or supportive (Article 6 TFEU). In an area of exclusive competence, 'only the Union may

---

[70] And further defined in Protocol (No. 2) on the application of the principles of subsidiarity and proportionality to the TFEU.

[71] For in-depth discussion please refer to K. Lenaerts and P. Van Nuffel, *European Union Law* (London: Sweet & Maxwell, 2013), at 112f.

[72] For a detailed discussion of the role of the subsidiarity principle, the requirements and the application in European law, please refer to ibid., at 131–40.

legislate and adopt legally binding acts' (Article 2(1) TFEU). By contrast, in the area of a shared competence, 'the Union and the Member States may legislate and adopt legally binding acts .... The Member States shall exercise their competence to the extent that the Union has not exercised its competence' (Article 2(2) TFEU). Articles 4 (2)(e) and (2)(i) TFEU mention the areas of 'environment' as well as 'energy' as an area of shared competence.

As De Búrca puts it, 'exercise of competences is seen to be linked to the question of their existence'.[73] Protocol No. 25 of the TFEU further clarifies that 'when the Union has taken action in a certain area, the scope of this exercise of competence only covers those elements governed by the Union act in question and therefore does not cover the whole area'. In addition, Declaration 18 in relation to the delimitation of competences clarifies that the Intergovernmental Conference (IGC) reserves the right to increase or reduce a competence conferred upon the Union.[74] These restrictions do not necessarily add additional legal value, but once more reflect the 'fear' of the Member States in transferring too many competences to the Union, they fear losing part of their sovereignty concerns which is especially present in the area of the shared energy competence, as previously illustrated above.[75]

The Recommendation, based on Article 292 TFEU with regard to the environmental and energy competence, refers to the Union's right to 'act in order to address environmental impacts and risks, as well as to clarify and/or complement existing environmental legislation'. However it acknowledges that 'choices of energy mix remain in the hands of Member States'.[76] Hence, with reference to the subsidiarity and proportionality principles, the Commission has to justify why its action on shale gas (an environmental and energy matter) is preferably regulated on a Union-wide basis.[77]

---

73 De Búrca, note 69 above, at 92.

74 Declaration 18 to the TEU in relation to the delimitation of competences.

75 See also H. B. Vedder, 'The Treaty of Lisbon and European environmental law and policy' *Journal of Environmental Law* 22(2) (2010), 285–99, at 292f; as well as O. Woodley, 'Governing a North Sea grid development: the need for a regional framework treaty' *Competition and Regulation in Network Industries* 14(1) (2013), 73–97, at 87f.

76 European Commission, 'Staff Working Document, Impact Assessment, Exploration and production of hydrocarbons (such as shale gas) using high volume, hydraulic fracturing in the EU', Brussels, 22.1.2014, SWD(2014) 21 at 39 ('Impact Assessment').

77 See sections 3.1.2 and 3.2 below.

As will be further discussed in section 4, the choice of the correct legal basis has practical implications. Article 191 TFEU on the environment includes the option for Member States to apply more stringent conditions than a Union measure proposes, the so-called 'environmental guarantee'. That means that the Member States generally have the possibility to establish stricter regulatory measures than included in the Union's legislation. For non-binding Recommendations outlining minimum principles, it is already implied that Member States can adopt stricter measures on shale gas; hence the environmental guarantee in the current state of affairs is not of great relevance, at least prima facie. To the extent that the Recommendation lists (what in the Commission's view are) binding obligations arising out of existing EU law, the environmental guarantee is evidently fully applicable. Furthermore, Article 194 TFEU asks on the one hand for solidarity and common actions of the Member States to achieve the energy and environmental objectives[78] while ensuring the security of energy supply in the Union, but on the other hand, the second paragraph leaves it to the Member States to exploit their resources and to choose their internal supply structure.

Once the institutions (mostly the Commission) have found 'a sufficient legal basis to act' (the 'power' to act),[79] it still has to be assessed 'as to whether it should act and, if so, to what extent. … ("if and in so far as")'.[80] The implications for shale gas are reviewed in the following section.

### 3.1.2 Subsidiarity and proportionality

The subsidiarity principle, without being referred to as such,[81] was originally introduced as a principle for environmental policy in the Single European Act.[82] Certain aspects of it were already included in the first

---

78 I. Pernice, 'The Treaty of Lisbon: multilevel constitutionalism in action' *Columbia Journal of European Law* 15 (2009), 349–407, at 49.

79 K. Lenaerts, 'The principle of subsidiarity and the environment in the European Union: keeping the balance of federalism' *Fordham International Law Journal* 17(4) (1993), 846–95, at 875.

80 Ibid., see also Commission Report to the European Council on the Adaptation of Community Legislation to the Subsidiarity Principle, COM (93) 545, at 7.

81 Lenaerts, note 79 above, at 859, as well as Schutze, note 69 above, at 173.

82 See also Lenaerts, note 79 above, at 859, as well as Schutze, note 69 above, at 173 and Van Calster and Deketelaere, note 12 above, at 12–25. For the history of the negotiations on the principle and its inclusion in the Treaty see J. Golub, 'Sovereignty and subsidiarity in EU enviornmental policy' XLIV *Political Studies* (1996), 686–703.

Environmental Action Programme. It was promoted under the Maastricht Treaty in Article 3b to become a general European law principle[83] for determining the competences between Member States and the Union,[84] or put differently the 'duty for Commission and Council to justify the exercise of competences'.[85] The overall aim of the subsidiarity principle is to ensure that power, including regulatory power, is exercised at the level where it will achieve the best results. The interpretation can hence go both ways: pro or contra more European involvement. The Lisbon Treaty added a new reference to the central, regional or local level for proposed actions, emphasizing the aim to determine measures at the best possible level of regulation: ensuring that these measures are adopted as close as possible to the citizens.[86] It thus follows the concept of decentralization, assuming that the most appropriate or the best level of regulations is the most local level. It requires and is based on interconnection, interdependence and cooperation between the Member States and the Union (a cooperative system of separation of powers), rather than the strict separation of powers as used in the US system.[87]

The subsidiarity principle imposes two criteria which both have to be met for an action at the Union level to be justified.[88] First, the 'objectives of the proposed action cannot be sufficiently achieved by [actions of] the Member States', and secondly, and simultaneously, they can be 'better' achieved at the Union level 'by reason of the scale or effects of the proposed action'.[89] A proposal for Union action in the area of the environment and energy always has to justify why the intended action

---

83 See also Lee, note 54 above, at 136.

84 For the history of the negotiations on the principle and its inclusion in the Treaty, see Golub, note 82 above.

85 A. Faludi, 'Territorial cohesion and subsidiarity under the European Union Treaties: a critique of the "territorialism" underlying' *Regional Studies* 47(9) (2012), 1594–606 at 1598.

86 Thus reflecting the 'proximity of government' see K. Lenaerts, note 79 above, at 865.

87 Bast and Bogdandy, note 68 above, at 236.

88 Interestingly, the UK in the past saw in the subsidiarity principle a means to limit EU power: Golub, note 82 above.

89 Article 5 Protocol No. 2 on the application of the principles of subsidiarity and proportionality. This is supported by the second paragraph of Article 5(3) TEU which refers to Protocol No. 2 to the TEU (originally adopted as Protocol No. 30 on the application of the Principles of Subsidiarity and Proportionality under the Amsterdam Treaty) and states that: 'The institutions of the Union shall apply the principle of subsidiarity as laid down in the Protocol on the application of the principles of subsidiarity and proportionality. National

will be more effective if taken at the Union level, and why actions at the national level are not sufficient to achieve a specific aim.[90] The notion of the word 'better' remains unclear: the Protocol on the application of the principles which is attached to the Lisbon Treaty[91] only indicates that it should be 'substantiated by qualitative and, wherever possible, quantitative indicators'.[92] However, as de Sadeleer puts it, this can mean 'more effective, more consistent, more democratic, more consistent with the internal market obligations, etc.'.[93]

Generally, measures in the area of environmental law and policy are said to pass the subsidiarity hurdle 'quite easily'.[94] The Commission generally has to justify why the powers and freedom of action of the Member States are limited in order to achieve the result intended.[95] However, it is also 'for the European Union legislature to determine the measures which it considers necessary to achieve the intended objectives, while observing the principles of subsidiarity and proportionality'.[96] Prominent arguments for European involvement in environmental issues are the often transboundary nature of the environmental problem(s); the overall aim of a high level of environmental protection and the argument that unilateral measures in the environmental area lead to a distortion of competition and thus contravene the internal market objectives.[97] In this regard, Lenaerts concluded in 1993 that 'subsidiarity will not stand in the way of the further development of Community environmental policy along the lines that it has been following so far'.[98]

---

Parliaments ensure compliance with the principle of subsidiarity in accordance with the procedure set out in that Protocol.'

[90] In the case of the Recommendation on shale gas, the section on 'the EU's right to act and justification' is included in the Impact Assessment, note 76 above at 39f. See also further below.

[91] On the history of the Protocol see: P. Craig, 'Competence: clarity, conferral, containment and consideration' *European Law Review* 29(3) (2004), 323–44 at 342.

[92] Article 5 Protocol No. 2. See on the notion of efficiency and effectiveness in determining the need for Union action: Lenaerts, note 79 above, at 877.

[93] N. de Sadeleer, 'Principle of subsidiarity and the EU environmental policy' *Journal for European Environmental and Planning Law* 9(1) (2012), 63–70, at 64.

[94] Ibid., 64f.

[95] ECJ, Case T-374/04 *Germany v Commission*, [2007] ECR II, 4431, at 79.

[96] ECJ, Case C-504/09 P *Commission v Poland*, [2012] ECR, nyr, at 79.

[97] Ibid.

[98] Lenaerts note 79 above, at 852.

With regard to energy policy, the application of the subsidiarity principle has not, as yet, passed the test in practice. Whereas in the environmental area the principle is used as a means to promote a European measure, it is expected to have the opposite effect regarding energy policy, and to limit the European action and integration in this policy area,[99] which is again due to the Member States' concern regarding their sovereignty over energy resources.[100]

The Commission indeed justified its action on shale gas, on grounds of its transboundary impacts. These result from the geology of the shales (seven European shale basins are situated in two or more Member States) and the fact that '[e]nvironmental risks and impacts may *spread across national borders*' (original emphasis).[101] The Impact Assessment refers to the specific transboundary risks for surface and groundwater (at minimum 268 transboundary water bodies), air quality and greenhouse gas (GHG) emissions.[102] From the Commission's point of view, Union action is thus needed and justified in the area of shale gas.

The principle of subsidiarity has to be read in conjunction with the principle of proportionality.[103] Whereas the subsidiarity principle relates to the necessity of a Union measure, the proportionality principle determines the intensity of a measure. Paragraph 4 of Article 5 TEU states that 'under the principle of proportionality, the content and form of Union action shall not exceed what is necessary to achieve the objectives of the Treaties'. The three underlying aspects are: (a) the sustainability or appropriateness between the means and the end of a measure; (b) the necessity of the action to achieve the proposed Union objective (in the area of the environment, the objectives under Article 191 TFEU); and (c) the proportionality of the measure itself (*sensu stricto*), meaning whether the measure is disproportionate with respect to the goal intended.[104] Article 5 of the Protocol further establishes that among other

---

[99] Franz and Kane, note 58 above, at 474.

[100] Vedder, note 75 above, at 292f, as well as Woodley, note 75 above, at 87f.

[101] Impact Assessment, note 76 above at 39. Recourse to the subsidiarity principle is also taken on pp. 24 and 42 of the Impact Assessment.

[102] Ibid.

[103] For a detailed discussion of the role of the proportionality principle, the requirements and the application in European law, please refer to Lenaerts and Van Nuffel, note 71 above, at 141–7, as well as Van Calster and Reins, note 11 above.

[104] These are stages used commonly in European law, see also, P. Craig and G. De Búrca, *EU Law: Text, Cases and Materials* (Oxford: Oxford University Press, 2003), at 372, as well as T. Harbo, 'The function of the proportionality principle in EU law' *European Law Journal* 16(2) (2010), 158–85, at 165; and

factors, the financial impact of a proposed legislative act needs to be taken into account. The Recommendation on shale gas is not a legislative act as such. However, the Impact Assessment does contain an 'EU Value Added Test'. This examines the proportionality of the measure, referring to an earlier Commission study which had concluded that addressing security supply concerns solely at the national level results in welfare losses, and that a Union approach would 'make the economic case of shale gas clearer'.[105] The Commission further refers to lessons learned in the offshore fossil fuels extraction sector, which was until the establishment of the Offshore Safety Directive, an unharmonized activity associated with liability, emergency planning and compliance concerns. The Commission further refers to the US's experience with shale gas.[106]

The Commission thus acts on the basis of Article 292 TFEU with respect to the shared environmental and energy competences. It does so on the grounds of transboundary environmental issues arising from the exploration and production of shale gas, the expected added value shale gas would bring to European energy supply security, as well as increased legal clarity, predictability and a 'level playing field for its safe and secure extraction'.[107] However, it is worth keeping in mind that a Recommendation as such is not a legislative proposal, and that, as a consequence, only a 'light' subsidiarity test as included in the Impact Assessment was carried out. The thresholds for a subsidiarity and proportionality analysis for legislative acts are higher. The Protocol provides for such acts a consultation of the national parliaments.[108] As Craig puts it, it is 'clear that subsidiarity has had an impact on the existence and form of Community [now Union] action. … The idea of Community [now Union] control with a "lighter touch" fits with changes that predate the Constitution [Lisbon Treaty].'[109] The following section will examine this issue and explain why the Commission chose the form of a non-binding Recommendation to address the issue of shale gas within the Union.

---

S. Lierman, 'Law as a complex adaptive system: the importance of convergence in a multi-layered legal order' *Maastricht Journal of European and Comparative Law* 21(4) (2014), 611–29, at 629.

[105] Impact Assessment, note 76 above at 39f.

[106] Ibid., at 40.

[107] Ibid.

[108] Article 4 of the Protocol No. 2, see further below.

[109] Craig, note 91 above, at 343.

## 3.2 Why Soft Law?

The Commission acted on the basis of the EU's shared competence in the areas of environmental and energy policy. As concluded above, the subsidiarity hurdle in the environmental area is generally easy to pass. The case of shale gas does not completely fit in the classical internal market environmental competence. In the past, measures affecting energy issues were also adopted in accordance with both of these competences.[110] The Lisbon Treaty, with the introduction of a specific competence on energy issues in Article 194 TFEU, changed the Union's power in this regard by creating a clear division of competences between the EU and the Member States,[111] providing for more transparency within energy issues.[112] Whether the competence set-up is really that clear will be examined in the next section. The inclusion of Article 194 TFEU has been hailed as a 'proper policy enabling clause'[113] for a European energy policy, showing that 'energy policy has evolved more than ever into a European subject and is – generally speaking – no longer to be dealt with as a matter for Member States only';[114] nor does it have to be adopted on other legal bases such as the internal market, competition or environmental competence.[115]

### 3.2.1 Article 194 TFEU as a game changer

As discussed in section 2.3 above, in the past energy issues were either regulated under environmental or internal market competences. However, with the new set-up in the TFEU, the environmental challenge presents itself here with an additional layer of complexity: ensuring that not just the *environmental* objectives of the Treaty are met, but also the *energy* policy objectives. Energy and environmental issues are no longer regulated under one basis.[116] The separate energy title gives energy issues (arguably) a stronger standing within the European system.

---

110 See also Vedder note 75 above, at 292f.

111 Bogdanowicz, note 50 above, at 202.

112 Nowak, note 51 at 53.

113 Pielow and Lewendel, note 45 above, at 152.

114 Ibid., 148.

115 See also Vedder note 75 above, at at 292f.

116 Not all of them have been so before, see also section 2, as well as Lee, note 54 above, at 134.

However, even if Article 194 TFEU were an appropriate legal basis to promote these aims,[117] the Union is reluctant to establish binding measures on shale gas. This hesitation is not without reason. Article 194 TFEU establishes a specific competence regarding energy issues while at the same time introducing boundaries and limitations. Union energy policy aims at ensuring the security of energy supply in the Union, (Article 194(1)(b)), while at the same time the Article clearly states that '[Union] measures shall not affect a Member State's right to determine the conditions for exploiting its energy resources, its choice between different energy sources and the general structure of its energy supply.' In the case of shale gas, Article 194 TFEU can be understood as a real 'game changer'. The limitation introduced in Article 194(2) is respected by the Recommendation on shale gas, which states in its Preamble that 'Member States have the right to determine the conditions for exploiting their energy resources, as long as they respect the need to preserve, protect and improve the quality of the environment' (Recital (1)). It is thus with good reason that the Commission opted for a measure under the environmental and energy competence and drafted an environmentally oriented guidance. The environmental focus of the Recommendation is absolutely crucial, as a measure under the energy competence alone seems fairly impossible: the regulation of exploration and production would have clearly impacted the conditions for exploiting energy resources. In passing a non-binding, environmentally oriented guidance, the Commission chose the safe option, thereby avoiding discussions on the Union's competence to adopt binding measures on shale gas under either the environmental or energy competence.

Furthermore, normally the subsidiarity principle requires a broad consultation prior to the proposal of a legislative act and the forwarding of the draft to the national parliaments, as well as an assessment of the financial impact and the consideration of financial and administrative burdens.[118] However, this is not required for a Recommendation, as a Recommendation is not considered to be a legislative act. This is certainly one reason why the Commission adopted guidance on shale gas in the form of a Recommendation.[119] The likelihood that Member States

---

[117] Which it is not, as will be established in the section on the horizontal competences.

[118] Article 2 and 5 Protocol No. 2. See also Craig, note 91 above, at 324.

[119] Article 292 TFEU.

such as Poland and the UK (which are against European involvement) would protest against a legal measure on subsidiarity grounds is considered to be quite high.

Article 194 TFEU however, also refers to the 'spirit of solidarity between Member States'[120] in order to 'ensure security of energy supply in the Union'. This reference has been interpreted in the past as a 'corrective of the subsidiarity principle', in the sense that the aims listed in Article 194 TFEU cannot be adhered to at a solely national level, and cooperation and action at Union level is needed.[121] The EU Security of Gas Supply Regulation[122] supports this interpretation. Drafted as a reaction to the 2006 and 2009 gas crises and the lack of solidarity between Member States, it understands the responsibility for security of gas supply as 'a shared responsibility of natural gas undertakings, Member States, ... and the Commission, within their respective areas of activities and competence. Such shared responsibility requires a high degree of cooperation between them.'[123] Cooperation and coordination in the area of shale gas extraction was a topic of discussion at the Informal Environmental Council meeting on 16 July 2013, and the Technical Working Group meetings organized for drafting the shale gas impact assessment.[124] While Member States revealed different views on the form of the European involvement on the issue, they were generally in favor of a common EU approach, with only one Member State articulating its opposition to any EU approach.[125] As reported earlier, this tendency has been proven in practice.

An additional argument has been put forward by Nettesheim.[126] Article 194(1)(c) TFEU relies on the Member States' solidarity to promote 'energy efficiency and energy saving and the development of *new and*

---

[120] See for a more detailed discussion section 5.4 below.

[121] Pielow and Lewendel, note 45 above, at 153; a second way of understanding the principle is to see it as a restriction of a European energy policy in the way that the Union is only competent to act under the energy competence to render the Member States more independent, as argued in P. Thieffry, 'Les politiques européennes de l'energie et de l'environnement: rivales ou allies?' *Revue des Affaires Européennes* 4 (2011), 783–809, at 791.

[122] Regulation No. 994/2010 of 20 October 2010 concerning measures to safeguard security of gas supply and repealing Council Directive 2004/67/EC, [2010] OJ L 295/22 ('Security of Gas Supply Regulation').

[123] Article 3 Security of Gas Supply Regulation.

[124] Impact Assessment, note 76 above at 19f.

[125] Ibid. at 19.

[126] M. Nettesheim, 'Das Energiekapitel im Vertrag von Lissabon' *Juristen-Zeitung* 65(1) (2010), 25–32.

*renewable* forms of energy' (emphasis added). According to Nettesheim, the words 'new' and 'renewable' are not necessarily cumulative, as they are separated by the connector 'and'.[127] Following his interpretation, it would mean that the Article also covers the promotion and development of new forms of energy such as shale gas.[128] Applying a narrow interpretation of that provision, one might argue that this provision ('the development of *new* … forms of energy') requires the Union to assess the potential of shale gas resources and its development. This would be in line with the policy approach the Commission is taking so far. Whether shale gas can be considered a new form of energy (technology) will be further debated in Chapter 3.

### 3.2.2 The use of soft law within the EU

A practical application of the proportionality principle arguably promotes the use of non-binding measures. The European Council noted in 1992 on the application of the subsidiarity and proportionality principles that:

> The form of action should be as simple as possible, consistent with satisfactory achievement of the objective of the measure and the need for effective enforcement. … Other things being equal, directives should be preferred to regulations and framework directives to detailed measures. Non-binding measures such as recommendations should be preferred where appropriate.[129]

Whether the Recommendation on shale gas can lead to a 'satisfactory achievement of the objective of the measure', thus ensuring that 'the public health, climate and environment are safeguarded, resources are used efficiently, and the public is informed' (section 1(1)), remains to be seen. The Commission will review the Recommendation's effectiveness 18 months after its publication, and 'decide whether it is necessary to put forward legislative proposals with legally-binding provisions on the exploration and production of hydrocarbons using high-volume hydraulic fracturing'.[130] The replies to the Questionnaire by the individual Member States are already published;[131] a final assessment has not yet been conducted.

---

[127] Ibid., at 20.

[128] Under weighting of subsidiarity issues of course, see also ibid.

[129] European Council Conclusions, Edinburgh, 12 December 1992, at 21. See also K. Lenaerts, note 79 above, at 885.

[130] Section 16.4 of the Recommendation.

[131] European Commission 'What's new?' available at http://ec.europa.eu/environment/integration/energy/unconventional_en.htm, last accessed 28.06.2015.

A problem regarding the use of soft law[132] is that '[v]oluntary programs lack incentives for "bad actors" ... to participate, thus leaving out the entities most in need of regulation'.[133] Whether this conclusion can be drawn for shale gas as well can only be established after the review of the application of the Recommendation by the Member States. However, looking at how Poland answered the Questionnaire,[134] it may well be the case that incentives for participation are lacking.

Nevertheless, one problem with the adoption of non-binding instruments, especially with Recommendations issued by the Commission (as opposed to Council Recommendations),[135] is the fact that they are not subject to democratic decision-making at European level and thus lack legal status despite remaining importance in practice.[136] In *Grimaldi,* the European Court of Justice (ECJ) turned to this question on the legal value of Recommendations observing that:

> The national courts are bound to take recommendations into consideration in order to decide disputes submitted to them, in particular where they cast light

---

[132] For a discussion of the legal implications of soft law and a brief comparison between hard and soft law, refer to M. Lee, *EU Environmental Law, Governance and Decision-Making*, 2nd ed. (Oxford and Portland: Hart Publishing, 2014), at 88.

[133] International Center for Technology Assessment, 'Principles for the oversight of nanotechnologies and nanomaterials' (2007), available at www.icta.org/doc/Principles%20for%20the%20Oversight%20of%20Nanotechnologies%20and%20Nanomaterials_final.pdf; last accessed 28.06.2015.

[134] Indicating 'partial' compliance and not providing additional and clarifying information. See also reply of Poland to the Questionnaire on the application of Commission Recommendation 2014/70/EU on minimum principles for the exploration and production of hydrocarbons (such as shale gas) using high-volume hydraulic fracturing, available at http://ec.europa.eu/environment/integration/energy/unconventional_en.htm; last accessed 30.06.2015.

[135] They arguably have some democratic legitimacy as they are adopted by the executive.

[136] See also L. A. J. Senden, 'Soft law and its implications for institutional balance in the EC' *Utrecht Law Review* 1(2) (2005), 77–99, at 80f, as well as B. A. Beijen, 'The implementation of European environmental directives: are problems caused by the quality of the directives?' *European Energy and Environmental Law Review* 20(4) (2001), 150–64, at 161; even if Beijen is not directly referring to Recommendations.

> on the interpretation of national measures adopted in order to implement them or where they are designed to supplement binding Community provisions.[137]

The Court further states that they 'cannot in themselves confer rights on individuals upon which the latter may rely before national courts'.[138] As such, the Recommendation in question does have legal effect in the sense that it can provide interpretational guidance in national court proceedings, but not more than that.[139]

## 4. THE HORIZONTAL COMPETENCES AND SHALE GAS

The relationship between the EU's energy and environmental competences with respect to renewable energy and its promotion and impact on climate change has been the subject of academic discussion[140] and judicial review.[141] The general tendency is for measures on renewable energy to be passed under both competences, depending on the specific content and objective,[142] following the *accessorium sequitur principale* rule: where the procedures are divergent, the act must be based on a single legal basis, namely that required by its main or predominant purpose or component, thus in this case either on Article 192 TFEU[143] on the environment or on Article 194 TFEU on energy

---

[137] Case C-322/88 *Grimaldi* [1989] ECR I-4407 at 18f.

[138] Ibid. at 19.

[139] See also C. G. Molyneux and L. Tosoni, 'What's in a name?: legal implications of the EU Recommendation on shale gas' *Environmental Law Institute* 44(5) 5 (2014), 10351–3. For a more detailed discussion on the use of soft law in the Union and its impact on shale gas policy, please refer to Chapter 3.

[140] See, e.g., Frenz and Kane, note 58 above, at 464.

[141] ECJ, Case C-2/10, *Azienda Agro-Zootecnica Franchini Sarl, Eolica di Altamura Srl v Regione Puglia*, [2011] ECR I-06561.

[142] Directive 2009/28/EC of 23 April 2009 on the promotion of the use of energy from renewable sources, [2009] OJ L140/16, for example, is based on the environmental and internal market competence; Directive 2012/27/EU of 25 October 2012 on Energy Efficiency, [2012] OJ L315/1 is based on the energy competence.

[143] Consolidated Version of the Treaty on the Functioning of the European Union, [2010] OJ C83/47 (TFEU).

(where legislative procedures are identical, the act may be based upon several provisions).[144] However, other forms of energy have not yet been included in this debate and the outcomes with respect to the exploration of unconventional gas are expected to be quite different, since such exploration is associated with negative environmental impacts, alongside the benefits it offers to the security of the EU's energy supply, which complicates the question of competence.

This section analyses the interplay and conflicts between the energy and environmental titles of the TFEU in the context of shale gas activities. The section further explores the concept of 'significance' in this setting, and how it is crucial in determining the appropriate decision-making procedure to pass a legal measure on shale gas in practice. Building on the analysis of the changes regarding the energy and environmental legal bases, this section explains the importance of choosing an appropriate legal basis for EU action and the procedure followed in doing so. The final part of the section considers the various possibilities for an appropriate legal basis for EU measures relating to shale gas.

## 4.1 Legal Basis for Shale Gas

### 4.1.1 The choice of the legal basis

The choice of the correct legal basis is generally important for a measure, and in this context is particularly of importance as Article 193 TFEU under the environmental competence includes the 'environmental guarantee', according to which Member States have the right to adopt more stringent provisions. For measures taken under the energy competence, Member States do not have this discretion. The legal basis further affects the decision-making procedure; for example, where unanimity is required (as it is in certain cases under the environmental competence); Member States have a veto right. As a consequence, measures adopted under unanimity are often vague, as they are necessarily compromises between all Member States.

The European courts frequently deal with the question of the correct legal basis of a measure.[145] From the Court's jurisprudence, four

---

144 G. Van Calster, 'Public environmental law in the European Union' in: R. Seerden, M.A. Heldeweg and K. Deketelaere (eds), *Public Environmental Law in the European Union and the United States* (Kluwer, 2002), 465–515, at 465; see also further below.

145 See, e.g., ECJ Cases C-300/89, *Commission v Council (Titanium Dioxide)*, [1991] ECR I-2867; C-155/91, *Commission v Council*, [1993] ECR

principles or criteria can be derived that are crucial when it comes to determining the correct legal basis of a measure regulating shale gas. First, the choice of the correct legal basis must be based on objective criteria amenable to judicial review.[146] Second, in order to establish the objective criteria, the ECJ generally employs the 'centre of gravity' or *accessorium sequitur principale* approach,[147] as determined in the landmark cases *Titanium Dioxide* and *Waste Directive*,[148] to decide which legal basis is the correct one in a given case. The main object and purpose of the proposed measure is what determines the legal basis that is used.[149]

Third, if two objectives have been identified, priority is given to the main purpose or main component[150] of the measure, thus on the predominant legal basis, as the Court concluded in *Commission v Council*[151] and emphasized for environmental issues in the *Huber* judgment.[152] Fourth, if the objectives and purposes of a measure are of similar importance, and 'indissolubly linked with each other',[153] a dual legal basis is possible,[154] unless there is a clear incompatibility of

---

I-939; C- 176/03, *Commission v Council*, [2005] ECR I-7879; and C-440/05, *Commission v Council*, [2007] ECR I-9097.

146 See ECJ, Case 45/86, *Commission v Council*, [1987] ECR I-1493, 1520, as well as Case C-338/01 *Commission v Council* [2004] ECR I4829, at 54, and Opinion 2/00 [2001] ECR I-9713, at 22

147 ECJ, Case C-300/89, note 145 above. See also M. Schemmel and B. de Regt, 'The European Court of Justice and the environmental protection policy of the European Community' *Boston College International and Comparative Law Review* 17(1) (1994), 53–83, at 53; R. van Ooik, 'Cross-pillar Litigation before the ECJ: Demarcation of Community and Union Competences' *European Constitutional Law Review* 4(3) (2008), 399–419, at 399.

148 ECJ, Case C-300/89, note 145 above; and ECJ, Case C-155/91, *Commission v Council (Waste Directive)*, [1993] ECR I-00939.

149 See ECJ Cases C-45/86, note 146 above; C-242/87, *Commission v Council*, [1989] ECR I-1425; C-155/91, note 148 above; C-376/98, *Germany v Parliament and Council*, [2000] ECR I-8419; C-176/03, note 145 above; C-94/03, *Commission v Council*, [2006] ECR I-1; and C-166/07, *Parliament v Council*, [2009] ECR I-7135. See also Lenaerts et al., note 71 above, at 118.

150 See Lenaerts et al., ibid.

151 ECJ, Case C-338/01, note 146 above.

152 ECJ, Case C-336/00, *Huber*, [2002] ECR I-7699. See also P. Wenneras, 'Towards an even greener union? Competence in the field of the environment and beyond' *Common Market Law Review* 45(6) (2008), 1645–85, at 1670ff.

153 ECJ, Case C-178/03, *Commission v Parliament and Council*, [2006] ECR I-107; and ECJ, Case C-155/07, *Parliament v Council*, [2008] ECR I-8103.

154 See also Lenaerts et al., note 71 above, at 119.

decision-making procedures.[155] If the decision-making procedures are incompatible, the 'more specific' Treaty provision, the 'provision that affords the appropriate legal basis by itself',[156] is crucial.[157] In case there is no Treaty provision 'supporting the whole of the intended measure', the measure has to be split into different parts.[158] These four general principles will be applied to the case of shale gas in the following section with the aim of establishing which Treaty provision is a suitable legal basis for a shale gas measure in the future.

### 4.1.2 First criterion: the objective and content of a measure

The first principle relates to the objective and content of a measure. As established in the introduction, shale gas is often cited as a solution to the increasing energy supply dependence of the EU on third countries such as Russia. Thus, it is often promoted as a welcome game changer in this situation, as it would reduce the EU's dependence on external energy sources and ensure its internal energy supply security. Hence, the objective of a measure on shale gas would clearly fall within Article 194 TFEU's objective of 'ensur[ing] security of energy supply in the Union'.

Further, it is in the interest of both individual Member States and the Union to ensure the (transboundary) protection of the environment. As established above,[159] shale gas is arguably associated with impacts on the environment, thus a measure would also clearly fall within the

---

155 As established in ECJ, Case C-300/89, note 145 above; Case C-338/01, note 151 above and Case C-178/03, note 153 above, among others. However, in Case C-166/07, *Parliament v Council ('International Fund for Ireland')* [2009] ECR I-07135, the Court held (at 69) that;

> when using two legal bases with different legislative procedures and when Article 352 TFEU is one of these, both procedures need to be complied with. That is to say in the specific case both the 'co-decision' procedure referred to in Article 251EC and the requirement that the Council should act unanimously.

156 See also Lenaerts et al., note 71 above, at 119.

157 ECJ, Case C-271/94, *Parliament v Council*, [1996] ECR I-1689. See also P. Craig and G. de Búrca, *EU Law: Text, Cases, and Materials*, 5th ed. (Oxford University Press, 2011), at 308.

158 See Lenaerts et al., note 71 above, at 119.

159 See, e.g., Forster and Perks, note 8 above; European Parliament, *Shale Gas: Opportunity or Threat to the Environment?* (2012), available at: www.europa.eu/pdfs/news/public/story/20120302STO39863/20120302STO39863_en.pdf, last accessed 26.06.2015; and J. Broderick, K. Anderson, R. Wood, P. Gilbert and M. Sharmina, *Shale Gas: An Updated Assessment of Environmental and Climate Change Impacts* (Tyndall Centre, 2011).

disrupted, the issue would be high on the political agenda with absolute priority, with no, or only very little, regard to environmental concerns.[253]

The updated European Energy Supply Security Strategy from May 2014 stresses that, 'during permitting and project implementation, due account should be taken of existing EU environmental legislation and guidance to ensure the environmental sustainability and secure public support and acceptance for a project'. The interconnection of the supply security and environmental protection concepts is rather weak. The use of the wording 'due account should be taken' is not very strong and does not include a clear and easily enforceable obligation. The requirement remains abstract, i.e., it does not provide indications of legislation.

In the strategy, the Commission further established eight key pillars[254] or areas of action, which are aimed at fostering closer cooperation between the Member States regarding energy supply security[255] under the scope of the solidarity principle.[256] The fifth[257] action area is of importance regarding shale gas. Under the heading of 'increasing energy production in the European Union', shale gas is considered to be able to 'partially compensate for declining conventional gas production'.[258] The Commission goes on to stress in the same sentence that this is only an option when 'issues of public acceptance and environmental impact are adequately addressed'.[259] This once again emphasizes the inseparable link between energy supply security and environmental protection, in the

---

253 Ibid.

254 These pillars relate to:

1. Immediate actions aimed at increasing the EU's capacity to overcome a major disruption during the winter 2014/2015;
2. Strengthening emergency/solidarity mechanisms including coordination of risk assessments and contingency plans; and protecting strategic infrastructure;
3. Moderating energy demand;
4. Building a well functioning and fully integrated internal market;
5. Increasing energy production in the European Union;
6. Further developing energy technologies;
7. Diversifying external supplies and related infrastructure;
8. Improving coordination of national energy policies and speaking with one voice in external energy policy.

255 European Commission, 'European Energy Security Strategy', Brussels, 28.5.2014, at 3.

256 Further discussed below.

257 European Commission, note 254 above, at 13.

258 Ibid.

259 Ibid.

form of energy production. In order to put this into practice the Commission suggests several key actions which need to be carried out before a 'commercial scale production' can become viable in the Union.[260] The notion of 'adequately' as included in the Commission's proposal suggests a statement of balance of the measures; the Commission however does not elaborate on what 'adequately' concretely entails. It did not specify whether the correct implementation of the Recommendation would pass this threshold.

Talus concludes that the European energy supply security system for regulation is no longer state-driven, but is, instead, market-based.[261] Energy policy is regulated through general energy market Directives (for example the Internal Gas Market Directive, but also measures on efficiency[262] and savings, such as regarding the Buildings Directive[263]) and security of supply instruments (Security of Gas Supply Regulation).[264] Consequently, the national energy markets become increasingly absorbed by a single European energy market.[265]

The former Security of Gas Supply Directive, which was initially based on reporting and coordination requirements, was updated in 2010 to a Regulation, which shifted more power to the EU level. The public sector, not the markets, carries the ultimate responsibility of ensuring

---

[260] The Commission launched 'a European science and technology Network on unconventional/hydrocarbon extraction' and 'organised an exchange of information between Member States, the relevant industries and non-governmental organisations promoting environmental protection in order to draw up Best Available Techniques (BAT) reference documents on hydrocarbons exploration and production (BREF)'. It similarly calls upon the Member States to 'include by carrying out Strategic Impact Assessments and setting up [a] one-stop shop for granting permitting procedures, in accordance with the Commission's guidance documents on Streamlining environmental assessment procedures for energy infrastructure and Projects of Common interest and on EIAs for large-scale trans-boundary projects' and to 'assess the potential of unconventional hydrocarbons taking full account of Recommendation 2014/70/EU in order to ensure that the highest environmental standards are implemented'. Ibid., 14.

[261] However, the influence of the state-driven policy remains stronger than originally anticipated.

[262] Directive 2012/27/EU of 25 October 2012 on energy efficiency, amending Directives 2009/125/EC and 2010/30/EU and repealing Directives 2004/8/EC and 2006/32/EC, [2012] OJ L 315/1.

[263] Directive 2010/31/EU of 19 May 2010 on the energy performance of buildings, [2010] OJ L 153/13.

[264] Talus, note 1 above, at 99.

[265] Ibid.

energy supply security.[266] The creation of a European approach shifted the responsibility away from the national authorities to the European Commission,[267] a process which needs to be continued so as to achieve a 'real' EU energy policy.[268]

When considering the definition of supply security within the Union, each criterion warrants special consideration when talking about shale gas extraction as a means to secure energy supply. To date, there are no commercial shale gas undertakings in the Union; even if Poland successfully fractured vertically,[269] a definitive timeline extraction on a commercial basis has not yet been provided. Furthermore, there is no indication other than the US's experience as a point of reference of how expensive or cheap commercial shale gas will be, nor are assessments regarding the availability of the technology to deliver yet completely reliable. Resource estimates differ considerably depending on the method of calculation used.[270]

The European Institutions,[271] as well as several authors,[272] stress another important dimension of the European perspective on supply security: the interdisciplinary or multidimensional nature of the issue. Accordingly, energy supply security is a 'multidimensional concept including external as well as internal action'.[273] It thus 'entails both domestic political affairs and multilateral economic aspects'.[274] Keeping the multidimensionality in mind, this volume focuses on shale gas as a potential means to change the internal supply security and the internal

---

[266] Ibid., 103.

[267] Ibid., 102f.

[268] Well aware of the fact that this is not what the Member States wish for and taking into account the sovereignty considerations as outlined above and that energy is a shared competence.

[269] See section 5.3.1 below.

[270] See for example differences in the estimation of the amount of resources in Poland: J. Teusch, *Shale Gas and the EU Internal Gas Market: Beyond the hype and hysteria* (CEPS Working Documents No. 369, 2012), at 3.

[271] Ibid.

[272] W. D'Haeseleer, 'Reflections on a Common European Energy Policy' *European Review of Energy Markets* 1(3) (2006), 1–21, at 10; as well as F. Baumann, 'Europe's way to energy security: the outer dimension of energy security: from power politics to energy governance' *European Foreign Affairs Review* 15 (2010), 77–95, at 87, and F. Baumann, 'Energy security as a multidimensional concept' 1 *CAP Policy Analysis*, (2008), 1–16; and Leal-Arcas and Filis, note 2 above, at 1229.

[273] Baumann (2008), ibid., at 4.

[274] Baumann (2010) note 272 at 87, as well as ibid.

energy policy, rather than on the external dimension. This narrow interpretation can be supported by the argumentation that Article 194(1) TFEU starts with the limitation, '[i]n the context of the establishment and functioning of the internal market and with regard for the need to preserve and improve the environment, Union policy on energy shall aim …'. It can, in theory, be argued that the fourfold objective seems to be limited to the internal market perspective and environmental perspective and consequently that Article 194 TFEU 'only provides [a] legal basis for "internal" security of supply measures and not for external ones'.[275] In addition, due to the multidimensionality of the concept it may well be that the EU has exclusive, shared or no competence at all, depending on the specific energy policy aspect involved (for example production, transport, efficiency or storage).[276]

Energy security is an important aspect in the quest for a CEP.[277] The Commission notes that the challenges for an energy policy relate to

---

[275] This would thus not be in line with the principle '*in foro interno in foro externo*', see also Article 3(2) TFEU. A different approach can be found in S. S. Haghighi, *Energy Security – The External Legal Relations of the European Union with Major Oil- and Gas-Supplying Countries* (Oxford: Hart Publishing, 2007) at 83f. It also could be argued that nothing in the provision limits the Union's competence, as it is for example the case for the new legal bases established under the Maastricht Treaty concerning culture (Article 167(4) TFEU), public health (Article 168(3) TFEU), research and technological development (Article 180 TFEU), education, youth and sports (Article 211 TFEU). In these provisions it is either expressly stated to 'foster cooperation with third countries and the competent international organizations' or that 'the Union … shall complete the activities carried out by the Member States' and 'within their respective spheres of competence'. The energy title on the other hand does not have such a reference, either for the cooperation, or for acting within its 'sphere of competence'. Following that, it can be assumed that there is an implied competence of the EU referring to the external dimension of energy issues, even if there is no explicit external competence mentioned.

[276] See also Leal-Arcas and Filis, note 2 above, at 1225, as well as section 3 and 4 on competences above.

[277] D'Haeseleer defines a common European energy policy as:

> A Common European Energy Policy should … be the roadmap for implementation of the mission made concrete by the strategic goals or objectives set by the EU authorities towards the sort of energy future desired for the Union … . A CEP must define measures for implementation, and as such, it must provide 'constraints or boundary conditions', on the one hand, and 'incentive', on the other hand, to satisfy the set strategic goals and objectives … .

sustainability, security of supply and competitiveness.[278] The Commission further states that '[t]he point of departure for a European energy policy is threefold: combating climate change, limiting the EU's external vulnerability to imported hydrocarbons, and promoting growth and jobs, thereby providing secure and affordable energy to consumers'.[279] Commercial extraction of shale gas would ease most of these parameters, and thus seems, at least in theory, to be the perfect candidate for realizing the Commission's goal.

## 5.3 Divergent Trends in Member State Policy on Shale Gas

Shale gas hydraulic fracturing has been undertaken on a commercial basis since 2006 in the US and has initiated not only a new 'energy revolution' in the country, but also is said to have had a 'radical impact on the global economy', as it has the potential to restructure and redraft the world map of energy exports and imports.[280] Other states, such as China, Algeria, Argentina and South Africa as well as some Member States in the EU are assessing the potential of the resource as a means to extract energy.[281] China, the second largest economy, plans to rely on shale gas (30 trillion $m^3$ of recoverable shale gas resources, 70 per cent

---

D'Haeseleer, note 272 above, at 1, as also cited in F. Goran, 'Improvements in the development of a common European Energy Policy in the years 2007–2011' *Journal of Comparative Politics* 6(1) (2013), 43–60, at 43. Even if it has been established before the Lisbon Treaty, thus before the existence of a specific Treaty provision for energy policy, the definition grasps the important nuances.

[278] See for example the former President of the European Commission J. M. Barroso, 'The security of Europe's energy supply: continuous adaptation' 22 *The European Files, Security of Energy Supply in Europe: Continuous Adaption,* (2011) at 7, as well as paper from the Commission/SG/HR for the European Council, An external policy to serve Europe's Energy Interests, S160/06 and European Commission, 'An Energy Policy for Europe', Brussels, 10.1.2007, COM(2007) 1.

[279] European Commission, ibid., at 5.

[280] A. Osborne and E. Gosden, 'US shale revolution will transform global economy, says BP', *The Telegraph*, 15.01.2014, available at www.telegraph.co.uk/finance/newsbysector/energy/oilandgas/10575331/US-shale-revolution-will-transform-global-economy-says-BP.html; last accessed 23.06.2015, as well as S. de Jong and W. Auping, *The Geopolitics of Shale Gas* (The Hague Center for Strategic Studies, 2014).

[281] See also B. Palmer and P. Shah, 'Recent developments in shale gas in the UK, Algeria, Argentina, Canada, China, India, Poland, South Africa, Ukraine and USA – an update' *Environmental Liability* 21(1) (2013), 14–24.

more than the US resources) in order to satisfy the estimated 25 per cent increase of energy consumption in the next five years. Through a joint venture with an American company, China intends to overcome technological and logistical hurdles.[282] This development is also important for the European gas market, as one of the main arguments for the call for shale gas in the Union is the need to stay competitive.

Prospecting and exploration is underway in several EU Member States, despite the activity not yet being pursued in a commercial manner. Although the EU is no longer listed among the top ten key players with technically recoverable resources globally (thus far behind China and the US), it has sizeable amounts of shale gas resources within its territory with an estimated 13,309 to 16,000 billion m.[283]

Among the EU Member States, there are currently three, possibly four trends regarding shale gas exploration and extraction: there are proponents such as the UK and Poland; opponents such as France and Bulgaria; and those remain 'undecided' such as Germany and the Czech Republic. A possible fourth category includes to States without potential such as Italy;[284] however, as these countries have not adopted a strategy on the issue, this category will not be discussed below. The challenge for a common European energy policy lies in aligning these different positions towards one common voice on shale gas in the Union.

---

[282] Global Energy Initiative, 'Global Energy Developments to Watch in 2014' *Global Energy Affairs* 2(1) (2014), 15–17, at 16.

[283] This was still the case in the International Energy Agency, *World Shale Gas Resources: An Initial Assessment of 14 Regions Outside the United States*, available at www.eia.gov/analysis/studies/worldshalegas/ last accessed 26.06.2015; no longer, however, in the updated version: there, the countries of China, Argentina, Algeria, the US, Canada, Mexico, Australia, South Africa, Russia and Brazil form the top ten; see also U.S. Energy Information Administration, *Technically Recoverable Shale Oil and Shale Gas Resources: An Assessment of 137 Shale Formations in 41 Countries Outside the United States* (2013), available at www.eia.gov/analysis/studies/worldshalegas/; last accessed 26.06.2015.

[284] See for example S. Jewkes, 'Italy says shale gas production never taken into consideration' *Reuters*, 5.11.2013, available at: www.rigzone.com/news/oil_gas/a/129961/Italy_Says_Shale_Gas_Production_Never_Taken_Into_Consideration#sthash.uRTOkUSZ.dpuf; last accessed 23.06.2015.

### 5.3.1 Emerging trends in proponent countries

Poland has been considered a 'test case for European shale gas development', which is going to determine the process in the EU.[285] According to the Polish Environment Ministry, 109 concessions for shale gas have been granted, which cover an area of 88,000 $km^2$ and 43 exploration wells.[286] In January 2014, Poland's leading role in shale gas exploration in the EU has been proved anew: the natural gas exploration and extraction company San Leon Energy announced the completion of flow testing in Poland's northern Baltic Basin in the Ordovician Shales. The company reports a production rate of up to 60,000 standard cubic feet per day, despite the well not yet being fully cleaned of fracturing fluids. Once fully cleaned up, a potential flow rate of 200,000 to 400,000 standard cubic feet per day is estimated. According to the Executive Chairman, it is 'the most encouraging vertical shale well test in Poland to date'.[287] This development is interesting as it shows that Poland is very serious about challenging the position of other Member States vis-à-vis shale gas.

In the UK, a temporary moratorium that was put in place after seismic events occurred in spring 2011 (during a drilling at the Preese Hall Site near Blackpool) was lifted in December 2012.[288] Since then, the UK together with Poland has been a pioneer in promoting the activity. For example, the UK Draft Finance Bill 2014 includes a new onshore or 'pad' allowance to 'support the early development of onshore oil and gas projects which are not commercially viable at the 62 per cent tax

---

[285] F. Meißner and D. Naumenko, *Non-Conventional Gas Regulation in Europe: Implications for Ukraine*, (Berlin/Kyiv: Institute for Economic Research and Policy Consulting, 2011), at 9.

[286] M. Wagrodzka, *Update on Hydrocarbon Law and Projects of Shale Gas Resources*, Department of Geology and Geological Concessions, available at www.usea.org/sites/default/files/event-/Shale_Gas_2_Ministry_of_Environment_Marta_Wagrodzka.pdf, last accessed 23.06.2015.

[287] San Leon Energy, 'Lewino-1G2: Successful Vertical Frac Leads to Horizontal Well', Press Release, 23.01.2014, available at www.sanleonenergy.com/media-centre/news-releases/2014/january/23/lewino-1g2-successful-vertical-frac-leads-to-horizontal-well.aspx; last accessed 23.06.2015. For more technical in depth information please also refer to the 'Technical Notes for Lewino-1G2, 23rd January 2014' supplementing the press release, available at www.sanleonenergy.com/media/947391/lewino-technical-paper-notes.pdf; last accessed 23.06.2015.

[288] DECC, 'Written Ministerial Statement by Edward Davey: Exploration for shale gas', available at www.gov.uk/government/news/written-ministerial-statement-by-edward-davey-exploration-for-shale-gas; last accessed 23.06.2015.

rate'.[289] The bill stresses that shale gas is an important resource to increase national energy security and will create thousands of jobs, generate investment and cut down energy prices, creating further incentives for the national industry, considering that both States depend today on energy deriving from coal mines.

In addition to taxation measures, the government has also published a Strategic Environmental Assessment (SEA)[290] under the SEA Directive[291] in preparation of the 14th round of onshore licensing taking place in 2015. The assessment identified significant effects of shale gas productions on local communities in comparison to the existing gas sector. The UK is the first European country to undertake a strategic environmental assessment of shale gas. The outcome does not include any surprising or new information, but confirms once more that the impact on the local communities is considerable. What remains slightly vague are the identified possible environmental impacts under the high-activity scenario; these are of prime concern to the public.

### 5.3.2 Emerging trends in opposing countries

Bulgaria and France imposed bans on the exploration and exploitation of hydrocarbons using hydraulic fracturing[292] in 2012[293] and 2011, respectively.[294]

In 2011 the authorities in Bulgaria granted an exploration and extraction licence for the northern part of the country. In reaction to massive

---

[289] HM Revenue and Customs and HM Treasury, 'Overview of legislation in draft' (2013), at A.58, available at www.gov.uk/government/uploads/system/uploads/attachment_data/file/264647/Overview_of_legislation_in_draft.pdf; last accessed 23.06.2015.

[290] DECC, AMEC, 'Strategic environmental assessment for further onshore oil and gas licensing environmental report' (2013), available at www.gov.uk/government/uploads/system/uploads/attachment_data/file/273997/DECC_SEA_Environmental_Report.pdf; last accessed 23.06.2015.

[291] Directive 2001/42/EC of 27 June 2001 on the assessment of the effects of certain plans and programmes on the environment, [2001] OJ L 197/30.

[292] See for a detailed discussion R. Fleming, 'Shale gas – a comparison of European moratoria' *European Energy and Environmental Law Review* 22(1) (2013), 12–32.

[293] See also Shale Gas Europe, 'Bulgaria', available at www.shalegas/-europe.eu/en/index.php/resources/shale-opportunities-in-europe/bulgaria; last accessed 23.06.2015.

[294] Act 2011-835 of 13 July 2011(Loi n° 2011-835 du 13 juillet 2011 visant à interdire l'exploration et l'exploitation des mines d'hydrocarbures liquides ou gazeux par fracturation hydraulique et à abroger les permis exclusifs de recherches comportant des projets ayant recours à cette technique).

public opposition the government not only withdrew the permit, but also imposed a ban prohibiting drilling with a pressure above 20 atmosphere pressure,[295] which resulted in a prohibition on drilling activities below 200 metres. The law was so restrictive that in practice it blocked all oil and natural gas exploration and extraction activities in the country.[296] In June 2012 the Bulgarian Parliament amended the law and lifted the 20 atmosphere pressure threshold, allowing other oil and gas activities to recommence, while prohibiting any hydraulic fracturing activities.[297]

In France, the authorities, by taking recourse to the precautionary principle among others, banned fracturing.[298] The US company Schuepbach Energy challenged the ban before the lower administrative court at Cergy-Pontoise. The case went to the French Constitutional Court which upheld the ban.[299] The company previously held two exploration licences in the south of France which were suspended after the ban in 2011.[300]

The positions of these two Member States represent a minority opinion in the EU. Bulgaria is one of the few Eastern European Member States which is not in favour of shale gas as an opportunity to reduce its energy dependence on Russia. However, with the rigid position taken by these Member States the quest for a CEP faces an extra hurdle.

### 5.3.3 Emerging trends in 'undecided' countries

Other Member States conversely have taken a more moderate stance. The Czech Republic and Germany are among the Member States that have

---

[295] Decision of the Bulgarian Parliament from 18 January 2012, available at www.parliament.bg/bg/desision/ID/13824; last accessed 23.06.2015.

[296] See K. Trichkova, *Regulatory Provisions Governing Key Aspects of Unconventional Gas Extraction in Bulgaria* (Milieu Ltd., 2013).

[297] Decision of the Bulgarian Parliament from 14 June, 2012, available at www.parliament.bg/bg/desision/ID/14031; last accessed 23.06.2015.

[298] See also for a detailed discussion E. Cocciolo, 'A paradigm shift for shale gas regulation: fracturing the (current) sustainable development paradigm and avoiding systemic risks' available at https://wiki.uef.fi/display/energy/Shale+Gas%3A+Regulatory+Developments (password protected). However, the fact that France is highly dependent on nuclear energy and has a very strong nuclear lobby might be a reason as well.

[299] Le Conseil Constitutionnel, Décision n° 2013-346 QPC du 11 Octobre 2013, *Société Schuepbach Energy LLC*.

[300] See also G. Van Calster, 'Fracking litigation in France – Reference to the Constitutional Court', available at http://gavclaw.com/2013/07/29/fracking-litigation-in-france-reference-to-the-constitutional-court/; last accessed 23.06.2015.

not yet adopted a final decision on their shale gas policy. In 2012, the Czech Environmental Minister annulled all existing licences and ordered a review of the applications as well as an investigation into a moratorium on shale gas exploration and extraction.[301] It remains to be seen how the government will finally move forward on the issue.[302]

In Germany the issue of shale gas hydraulic fracturing has been on the political agenda since 2012. Initially, several 'Federal' States ('Länder'), such as North Rhine-Westphalia, imposed a moratorium on the activities until a federal decision on shale gas was reached,[303] on grounds of the scientific uncertainties and lack of knowledge regarding the exact impact on environmental resources.[304] After the elections of September 2013, the new Grand Coalition included a section on '*Fracking*' in its Coalition Agreement. The section stresses the 'enormous risk potential'[305] of the technology, and the fact that, as yet, the effects on humans, nature and the environment are not sufficiently scientifically verified. Accordingly, the approval of any shale gas activity can only be granted if the necessary data for the assessment is available, and only can be determined beyond all doubt that the technology does not have any adverse effect on water quality, and is thus in line with the 'duty of care principle' ('*Besorgnisgrundsatz*') of the Federal Water Act. The coalition will, with the involvement of the Länder, the scientific community and industry representatives, develop which specific findings the explorative undertakings have to provide to eliminate gaps in knowledge and to provide a sufficient basis for possible subsequent steps in reviewing the policy approach taken and possibly allowing hydraulic fracturing.[306] The agreement does not include any new findings; it essentially confirms the state of the German practice and its careful attitude in this regard.

---

301 T. Dąborowski and J. Groszkowski, *Shale Gas in Bulgaria, the Czech Republic and Romania. Political Context – Legal Status – Outlook* (Warsaw: OSW Report, 2012), at 16.

302 See also on this issue, Fleming note 292 above, at 25.

303 Joint decree of the Ministries of Economics and Environment (V B 1 – 47-03/IV-5-3052-37727).

304 Umweltbundesamt, 'Einschätzung der Schiefergasförderung in Deutschland' (2011), at 19f.

305 '… eine Technologie mit erheblichem Risikopotenzial' CDU, CSU und SPD, 'Deutschlands Zukunft gestalten – Koalitionsvertrag zwischen CDU, CSU und SPD', 18. Legislaturperiode, at 44, available at www.cdu.de/sites/default/files/media/dokumente/koalitionsvertrag.pdf; last accessed 24.06.2015.

306 Ibid.

The different positions of the Member States on shale gas policy at the national level certainly influence law- and policy-making at the Union level.

## 5.4 Changing the Picture Again: Article 194 TFEU and the Solidarity Principle

The reason why the Commission decided to adopt non-binding principles on the issue and did not suggest a proposal for a Directive for example, is the fear of subsidiarity and proportionality discussions.[307] Energy issues are still very much a prime concern of the Member States' sovereignty. The new Article 194 TFEU also includes a reference to solidarity between the Member States, which was absent in the earlier version of the provision in the Draft Constitutional Treaty. This reference is of special importance in terms of the establishment of a Common Energy Policy.

The principle of solidarity is one of the founding principles and values of the EU, together with democracy and the rule of law.[308] The Schuman Declaration in 1950 stated that 'Europe will not be made all at once, or according to a single plan. It will be built through concrete achievements which first create a de facto solidarity.'[309] The principle is applicable in a multitude of subject areas of European concern, including energy policy.[310] On this subject the Preamble to the Treaty Establishing the European Coal and Steel Community Treaty (1951) acknowledged that 'Europe can be built only through real practical achievements which will first of all create real solidarity, and through the establishment of common bases for economic development.'[311] The Maastricht Treaty introduced in its Preamble the notion that Member States desire 'to deepen the solidarity between their peoples while respecting their history,

[307] See also Commission Staff Working Document, 'Impact Assessment, accompanying the Commission communication on the exploration and production of hydrocarbons (such as shale gas) using high volume hydraulic fracturing in the EU Commission', Brussels, 2014, at 37f.

[308] I. Hartwig and P. Nicolaides, 'Elusive solidarity in an enlarged European Union' *Eipascope* 3 (2003), at 19.

[309] The Schuman Declaration, 9 May 1950.

[310] See also S. Andoura, *Energy Solidarity in Europe: From Independance to Interdependance* (Notre Europe Jacques Delors Institute, 2013), at 18.

[311] A. Sangiovanni, 'Solidarity in the European Union' *Oxford Journal of Legal Studies* 35(2) (2013), 1–29, at 1.

their culture and their traditions', an important sentiment which also requires special consideration regarding the Member States' energy supply concerns.[312]

There is no overall definition of the solidarity notion in the current Treaty nor has there been in previous ones. This is important as solidarity can mean different things to different Member States and governments, especially when it comes to energy solidarity obligations. Generally, solidarity can be seen as a 'principle which distinguishes the EU and its members from other parts of the world and international organisations'.[313]

The TFEU recognized the principle of solidarity as a fundamental governance principle and includes references to it in several instances.[314] A reference to energy solidarity is included in two specific occasions in the Treaty, namely in Articles 122 and 194 TFEU, as well as indirectly in Declaration 35 on Article 194 TFEU. Accordingly, the aims prescribed in Article 194 TFEU shall be reached 'in a spirit of solidarity between Member States'. The so-called 'energy solidarity clause' has been introduced into the renegotiation of the failed Constitutional Treaty led by Poland and other Baltic States as a reaction to the 2006 energy supply crisis. While the solidarity concern applies to the achievement of all four energy policy aspects, a limitation on energy solidarity is introduced in Declaration 35. The Declaration states that the provisions do 'not affect the right of the Member States to take necessary measures to ensure energy supply under the conditions provided in Article 347', meaning that in the event of war or serious international tension, the Union's measures should not limit the Member States' power to secure energy supply themselves.[315]

In addition, Article 122(1) states that:

> Without prejudice to any other procedures provided for in the Treaties, the Council, on a proposal from the Commission, may decide, in a spirit of solidarity between Member States, upon the measures appropriate to the economic situation, in particular if severe difficulties arise in the supply of certain products, *notably in the area of energy.*

The Pre-Lisbon Article 100 EC Treaty did not include this new insertion ('notably in the area of energy'). It thus precisely reflects the concern

---

[312] See also Fischer, note 45 above, at 52.

[313] Hartwig and Nicolaides, note 308 above, at 19.

[314] Articles 2, 3 and 21 TEU and Articles 67, 80, 122, 194 and 222 TFEU, as well as Protocol 28, Declaration 37 and 62.

[315] See Haghighi, note 275 above, at 80.

about the Union's energy dependency on external suppliers and shortages of energy supply. The original aim of this clause was to enable the Union to be better equipped to send a strong message when Member States' energy supplies are threatened.[316]

What difference the solidarity clause makes in practice regarding shale gas is not clear: does solidarity mean finding a common voice on energy issues? One could interpret this new reference to solidarity on energy issues in the Treaty as a legal obligation of the Member States to do so. Otherwise solidarity could be interpreted as indifference, which would go against the very intent of the Founding Fathers and their interpretation of the concept. It would further contradict the aim of the European Energy Security Strategy, which includes the need for solidarity and cooperation as one of the main tools in order to enhance energy security in the entire Union.[317] Due to the interdependence of the Member States,[318] decisions in one Member State can affect the energy security of another.[319] This interpretation of the notion clashes with the provision in Article 194(2) TFEU on the Member States' freedom to choose their own energy resources. As it turned out, at least regarding shale gas, the new reference to energy solidarity in the Treaty did not initiate a move towards a Common Energy Policy. On the contrary, the Member States are far from speaking with a unified voice on this important matter. As such, it seems that the provision is rather an empty shell and does not provide any impact on 'turn[ing] this Energy Union into reality'[320] as suggested by Maroš Šefčovič.

## 6. CONCLUSION

This chapter has analysed the 'constitutional' issues surrounding the shale gas debate. It has illustrated the relationship between Title XX on environment and Title XXI on energy and their interaction after the

316 For reasons of completeness it should be mentioned that the solidarity clause in Article 222 TFEU, where the EU and its Member States can act jointly in solidarity in case of a terrorist attack or a natural or manmade disaster is also important when it comes to energy solidarity. This legal basis could be used in order to deduce a competence for the external dimension of energy issues especially.

317 European Commission, note 254 above, at 2, 3, 17 and 20 for instance.

318 Ibid., 3.

319 Ibid., 8.

320 Ibid.

Treaty of Lisbon. The extent to which the Union can act on shale gas has been analysed; per the non-binding Recommendation, it has been concluded that the Commission is already acting at the brink of its competences. The Commission itself promoted the creation of a level playing field as one of the main objectives of its Recommendation. Despite being well drafted, the format contradicts the Commission's own aim, although the non-binding format might be the only way to act at all.

Through the non-binding Recommendation the EU institutions, contrary to what had been asked for by many non-governmental organizations, politicians, legal experts and citizen initiatives, did not adopt a legally binding (proposal for a) Directive, but opted for minimum principles for shale gas activities.[321] Such a Recommendation does not provide new ammunition for either proponents or opponents of the activity. While the minimum principles on shale gas are an important indicator of European involvement on the issue, the fact that these are non-binding also illustrates the reluctance and possibly the anxiety of the EU Institutions to interfere with an issue which has in the past been recognized as a key component of Member States' sovereignty. The Commission has no exclusive competence in the environmental and energy areas, and thus has to justify its action on shale gas. Overall, shale gas regulation may arguably pass the subsidiarity hurdle on transboundary environmental grounds with ease, although the limitations introduced in Article 194(2) TFEU can just as easily prevent a European (legal) measure on shale gas. The reasons why the Commission acted the way it did are in essence twofold: first, the fact that Article 194 TFEU functions as a 'game changer' in the case of shale gas and prevents a European measure from being implemented in its entirety, and secondly, the fear of 'real' subsidiarity discussions if a legislative proposal on shale gas has to go through national parliaments.

Contrary to what one would expect, legal measures under the energy competence concerning the exploration and extraction of shale gas resources are not possible; the environmental competence is the only possible legal basis for such measures. Furthermore, policy-makers in the European Institutions are expected to argue from the start that a measure

[321] DG for Internal Policies, Policy Department A: Economic and scientific policy, Impacts of shale gas and shale oil extraction on the environment and on human health (2011), for example came to the conclusion that 'consideration should be given to developing a new directive at the European level regulating all issues in this area comprehensively', at 9.

is needed under the environmental competence and not as a means to ensure energy supply, as such an argument focused on ensuring security of energy supply as an overarching aim would preclude the Union from acting in this regard. This constitutes a major departure from renewable energy measures, which can generally be enacted under both competences and are categorized by the specific content of each measure.

Additionally, if a shale gas measure in the form of a Directive or Regulation were to be adopted under the environmental competence, it is essential to determine whether these shale gas measures would '*significantly* affect' or merely 'affect' the choice between energy sources and general structure of energy supply. If it would merely 'affect' the components, a measure would be subject to the ordinary legislative procedure. However, if it would *significantly* affect the components, it would most likely be subject to unanimity and thus subject to veto rights from the Member States. If this were to be the case the likelihood that Member States such as Poland and the UK, which are generally quite keen on shale gas exploration and fear European intervention, would block such a measure in its entirety is considered to be high. A measure on shale gas at the EU level would consequently fail. Therefore the only way to pass a Union measure in the form of a Directive or Regulation on shale gas would be under the environment title without a significant effect being determined. Hence, under the condition that the ordinary legislative procedure is applied. The fact that there has never been a Directive adopted under the second paragraph of Article 192 TFEU further illustrates this complexity. Whether the EU legislature manages to jump over this relatively high hurdle remains to be seen. In addition, the threshold of 'significance' and the role the court might have to play in determining the value of the notion in practice might render regulation unpredictable. The Court becomes the '*bouche de la loi*', the unpredictability of the notion of the word 'significance' in this regard gives too much power to the Court. Legal systems should be able to deal with this uncertainty. One could argue that the reference to unanimity for specific environmental and energy measures in Article 192 TFEU should be deleted; the Treaty therewith amended because of a lack of application of the measure. In practice, legislative drafts are continuously watered down so as not to trigger the unanimity requirement, thereby being exposed to the blocking of a measure in its entirety.

Regarding a CEP, there is a plethora of policy documents on the development of European energy policy after 2005, and especially after 2009, which paves the way towards a common energy policy in the future. Taking into account the recent emergence of shale gas technology, two scenarios are possible. First, it is possible that proponent countries, such as Poland and the UK use the existing strategies and roadmaps as an anchor point to promote and reason for their engagement in shale gas activities, for instance regarding the potential of the activity to reduce greenhouse gas emissions. Second, conversely, it is equally likely that opposing countries use the same policy documents as a basis for an argument against shale gas on grounds of environmental concerns and knowledge gaps. The current existing policy framework is thus interpretable both ways, for and against a European voice on the issue.

None of the strategies includes a strong recognition of shale gas as a possible game changer. Considering the strong and very divergent positions the Member States have on shale gas, as well as the approach taken by the Commission (Recommendation and not a proposal for a Directive), it is questionable whether the current path towards a common policy can be continued upon in the future. From today's point of view it seems possible that Member States, driven by the possibility to gain energy independence, will rely on their competence to determine their energy mix and pursue their own policy independent of pre-existing roadmaps and strategies.

Nevertheless, a common approach is needed. Issues at stake are transboundary impacts due to the geography of the Union shales, as well as the need for a common voice against competition from the US and China. The division and lack of dialogue on the issue prevent sound environmental regulation of the activity. In this regard the Commission has already taken important steps through the various studies and working groups which resulted in the establishment of minimum principles. The Union Institutions, especially the Commission, have taken the lead on this issue and continue the exchange of information, best practice examples and public information.

# 3. Regulation of technology versus technology of regulation – risk and regulatory design of 'new' technologies in the European Union – lessons for and from shale gas

## 1. INTRODUCTION

Although the case study of shale gas is the first in which the new set-up of environmental and energy competences as introduced in the Lisbon Treaty plays out in practice, the regulation and balancing of energy and environmental matters in a coherent way has always been a key challenge for lawmakers[1] – especially in relation to so called 'new' technologies.[2] In order to establish how regulatory design and policy impact the development of a technology, the current procedure of dealing with

---

[1] See for example regarding wind energy and nature protection: R. Frins and H. Schoukens, 'Balancing wind energy and nature protection: from policy instrument towards genuine sustainable development?' in: L. Squintani and H. H. B. Vedder (eds), *Sustainable Energy United in Diversity: Challenges and Approaches in Energy Transition in the EU* (European Environmental Law Forum, 2014), 85–110.

[2] Especially regarding the areas of climate, air, land, biodiversity, waste or water. See European Commission, DG Environment, 'Energy and environment', available at http://ec.europa.eu/environment/integration/energy/index_en.htm; last accessed 24.06.2015; see also: E. Stokes, 'Recombinant regulation and the role of expertise: the case of synthetic biology' presented at the ACELG Workshop on Regulating risks in the European Union – The co-production of expert and executive power, Amsterdam, 21–22.05.2015, as well as R. Brownsword and M. Goodwin, *Law and the Technologies of the Twenty-First Century: Text and Materials* (Cambridge: Cambridge University Press, 2012) and ibid.

cross-cutting energy and environmental issues needs to be examined.[3] More precisely, the regulation of carbon capture and storage (CCS) and nanotechnology will be studied, as these issues also affect several elements of the environment and have posed regulatory challenges to the European Institutions in the past, prior to the inclusion of a separate legal basis for energy.[4]

While all these technologies were – and to some extent still are – perceived to be new, the regulatory approaches to them differ. The regulatory design debate regarding nanotechnology was, at least in the beginning, characterized by a cooperative approach between industry, non-governmental organizations and the general public.[5] Action at a European level includes a European strategy for nanotechnology,[6] a European action plan,[7] a common definition established though a Recommendation[8] and a code of conduct.[9] However, there is no overall legislation on this matter. The general health, safety and environmental legislation, such as the REACH Regulation,[10] the Plant Protection Directive, and the so-called 'new approach on technical harmonization and standardization' legislation (especially regarding machinery, personal

[3] See also on the importance of comparing different technologies and the regulatory intervention: L. B. Moses, 'Why have a theory of law and technological change?' *Minnesota Journal of Law, Science & Technology* 8(2) (2007), 589–606, at 597.

[4] This chapter is partly based on L. Reins, 'European minimum principles for shale gas: preliminary insights with reference to the precautionary principle' *Environmental Liability,* 22(1) (2014), 16–27.

[5] The environmental NGO Environmental Defense Fund and DuPont launched in 2007 the 'Nano Risk Framework' initiative which developed a risk assessment tool. For more information on the initiative see http://business.edf.org/projects/featured/past-projects/dupont-safer-nanotech/; last accessed 24.06.2015.

[6] European Commission, 'Towards a European strategy for nanotechnology', Brussels, 12.5.2004, COM(2004) 338.

[7] European Commission, 'Nanosciences and nanotechnologies: an Action Plan for Europe 2005–2009' (COM(2005) 243.

[8] Recommendation of 18 October 2011 on the definition of nanomaterial, [2011] OJ L 275/38.

[9] Recommendation of 7 February 2008 on a Code of Conduct for responsible nanosciences and nanotechnologies research, C(2008) 424.

[10] Regulation No. 1907/2006 of 18 December 2006 concerning the Registration, Evaluation, Authorisation and Restriction of Chemicals (REACH), establishing a European Chemicals Agency, amending certain Directives, [2006] OJ L 396/1.

protective equipment, medical devices, etc.) and more sector-specific regulation on for example cars, medical products, food or cosmetics applies.

The discussion surrounding the CCS technology resulted in the adoption of binding legislation at a very early stage of technological development.[11] The CCS Directive 2009/31/EC amends the Water Framework Directive,[12] the Waste Directive and several other Directives.[13] The European Commission itself promotes that 'although the components of CCS are all known and deployed on a commercial scale, integrated systems are new, and a clear regulatory framework is required. The EU's CCS Directive provides this.'[14]

The shale gas debate in the EU is characterized by the classical between industry and NGOs. A non-binding Recommendation outlining minimum principles is the only specific regulatory instrument adopted so far, aside from the general regulatory framework for conventional gas. When analysing and comparing these legal regimes and the legislative histories thereof, important conclusions and lessons can be learned for the regulation of the shale gas technology. Shale gas is considered to be

---

[11] See also I. Havercroft and R. Macrory, 'Pulling the threads together' in: I. Havercroft, R. Macrory and R. B. Stewart (eds) *Carbon Capture and Storage: Emerging Legal and Regulatory Issues* (London: Hart Publishing, 2011), 277–96, at 303, as well as R. Purdy and I. Havercroft, 'Carbon capture and storage: developments under European Union and international law' *Journal for European Environmental & Planning Law* 4(5) (2007), 353–66, at 366.

[12] Directive 2000/60/ of 23 October 2000 establishing a framework for Community action in the field of water policy, [2000] OJ L 327, at 1–73.

[13] See also K. J. De Graaf and J. H. Jans, 'Environmental law and CCS in the EU and the impact on the Netherlands' in: M. M. Roggenkamp and E. Woerdman (eds) *Legal Design of Carbon Capture and Storage: Developments in the Netherlands from an International and EU Perspective* (Antwerp: Intersentia, 2009), 157–81, at 158f.

[14] See European Commission, DG CLIMA, 'Ensuring safe and environmentally sound CCS', available at http://ec.europa.eu/clima/policies/lowcarbon/ccs/index_en.htm; last accessed 26.06.2015. It has also been referred to as the 'first comprehensive CCS legislation': D. Langlet, 'Transboundary dimensions of CCS: EU law problems and prospects' *Carbon and Climate Law Review* 8(3) (2014), 198–207, at 198. If this is mirrored at a Member State level is assessed for the Netherlands in De Graaf and Jans, note 13 above, at 178ff. They are of the opinion that a 'suitable balance has been found between existing environmental law and the wish to stimulate CCS activities in the near future in an environmentally safe way' (at 198) and thus uphold the Commission's statement.

a 'new'[15] and 'risky'[16] technology. However, one can question whether the technology are really new, and whether uncertainties are associated with the activity or whether it is rather the likelihood of the impacts which are in question.

This chapter analyses these trends in EU regulation 'new' technologies. It compares the regulatory approach taken on shale gas, with the approach taken for nanotechnology and CCS. Two questions are addressed:

- What are the lessons learned from regulatory design (failures) of technologies of the past?
- What is the 'new' in new technologies? And how does this matter for the regulation?

The chapter addresses the role of science in new technology regulation, then focuses on two dimensions which are important when talking about a new technology:[17] first, the variable of 'time' and its impact on regulatory design (the 'when'-dimension), and secondly, the fact that the perception of risks and uncertainties associated with a technology determines its regulatory framework (the 'risk'-dimension).[18] Finally, the regulatory approaches taken for CCS and nanotechnology will be included in the analysis and compared to the approach taken in respect of the shale gas technology.

---

[15] See for example: A. Huc and R. Vially, 'New perspectives for fossil fuels: hydrocarbons in "unconventional" settings' in: J. B. Saulnier and M. D. Varella (eds), *Global Change, Energy Issues and Regulation Policies* (Dordrecht: Springer, 2013), 47–75.

[16] M. LaBelle and A. Goldthau, 'Escaping the valley of death? Comparing shale gas technology policy prospects to nuclear and solar in Europe' *The Journal of World Energy Law & Business* 7(2) (2014), 93–111.

[17] See also G. Laurie, S. H. Harmon and F. Arzuaga, 'Foresighting futures: law, new technologies, and the challenges of regulating for uncertainty' *Law, Innovation and Technology* 4(1) (2012), 1–33, at 13.

[18] See also L. B. Moses, 'How to think about law, regulation and technology: problems with "technology" as a regulatory target' *Law, Innovation and Technology* 5(1) (2013), 1–20, at 11.

## 2. PROBLEM STATEMENT: LAW, SCIENCE AND NEW TECHNOLOGIES

In the last four decades alone, humans have created more scientific knowledge than in the previous 5,000 years.[19] Alongside the increase of scientific knowledge came the development of new technologies,[20] not only in the energy, environmental and health areas, but also in other sectors such as transportation, medicine, computer science and communication.[21] In academia, the emergence of new technologies over the years became a 'separate' discipline of study, often referred to as 'Science, Technology and Society' (STS), combining social science, law and engineering.[22]

In the legal discipline,[23] mainly in the international, American and European arenas, 'regulation' and its limitations have become the focus of attention.[24] This has been the case especially in combination with the

[19] J. Garreau, 'Science's mything links: as the boundaries of reality expand, our thinking seems to be going over the edge' *Washington Post* (23.07.2001), as cited in: G. E Marchant, 'The growing gap between emerging technologies and the law' in: G. E. Marchant, B. R. Allenby and J. R. Herkert (eds) *The Growing Gap between Emerging Technologies and Legal–Ethical Oversight* (Dordrecht: Springer, 2011), 19–33, at 20.

[20] For the different stages of the introduction of new technologies and associated challenges, refer to: A. Vedder and B. Custers, 'Whose responsibility is it anyway? Dealing with the consequences of new technologies' in: P. Sollie and M. Duwell (eds) *Evaluating New Technologies: Methodological Problems for the Ethical Assessment of Technology Developments* (Dordrecht: Springer, 2009), especially at 30f, as well as H. Jonas, 'Technology and responsibility: reflections on the new tasks of ethics' *Social Research* 40(1) (1973), 31–54, at 32.

[21] The invention of the telephone, the computer and the internet, to name just a few.

[22] See for example the European Association for the Study of Science and Technology (EASST) which was founded in 1981 and embodies science and technology study researchers, bringing 'together a variety of disciplines. Many of its members have qualifications in both natural science, engineering and social sciences.' See EASST, 'About EASST', available at http://easst.net/about-easst/; last accessed 25.06.2015.

[23] Which is already a complex system in itself, see also N. Luhmann, *Das Recht der Gesellschaft* (Suhrkamp Taschenbuch, 1995).

[24] Especially techno-scientific regulation, see also S. Lierman and L. Veuchelen, 'The optimisation approach of ALARA in nuclear practice: an early application of the precautionary principle? Scientific uncertainty versus legal uncertainty and its role in tort law' *European Environmental Law Review*

pace of technological development[25] and the fact that legislation is often a result of co-production between executive power and experts.[26] Concerns, criticism and legal research focused on the issue that regulation in place becomes quickly outdated.[27] Revision is not completed in a timely or efficient manner, and for many technologies, the legal framework is static rather than flexible.[28] Another point of concern is that the more

---

15(4) (2006), 98–107, at 98. For a list of problems associated with technological change, refer to L. B. Moses, 'Recurring dilemmas: the law's race to keep up with technological change' *University of Illinois Journal of Law, Technology & Policy* 2 (2007), 239–85, at 243.

[25] D. Sarewitz, 'Anticipatory governance of emerging technologies' in: G. E. Marchant, B. R. Allenby and J. R. Herkert (eds) *The Growing Gap between Emerging Technologies and Legal–Ethical Oversight* (Dordrecht: Springer, 2011), 95–105, at 95. Regulation of a technology focuses on the specific technology at a specific moment. See K. Tranter, 'The laws of technology and the technology of law' *Griffith Law Review* 20(4) (2011), 753–62, at 754. As well as Moses, note 3 above, at 597.

[26] S. Jasanoff, 'The idiom of co-production' in: S. Jasanoff (ed.), *States of Knowledge: The Co-Production of Science and Social Order* (Routledge 2004), 1–12, as well as E. Fisher, 'Chemicals as regulatory objects' *Review of European, Comparative and International Environmental Law* 23(2) (2014), 163–71, at 163. Oreskes stresses that in a 'perfect world' this indeed would not be the case and that 'scientists collect facts, politicians develop policies based on those facts, legislators pass laws to implement these policies, and government agencies enforce the laws, most likely through regulations based on the same kind of facts'. N. Oreskes, 'Science and public policy: what's proof got to do with it?' *Environmental Science & Policy* 7(5) (2004), 369–83, at 370. See also A. Woyke, 'The world view of nanotechnology – philosophical reflections' in: F. Jotterand (ed.) *Emerging Conceptual, Ethical and Policy Issues in Bionanotechnology* (Dordrecht: Springer, 2008), 13–26, at 18. See also D. Birnbacher, 'Emotions within the bounds of pure reason: emotionality and rationality in the acceptance of technological risks' in: S. Roeser (ed.) *Emotions and Risky Technologies* (Dordrecht: Springer, 2010), 177–94, at 177.

[27] See for example L. M. Gaudet and G. E. Marchant, 'Administrative law tools for more adaptive and responsive regulation' in: Marchant, Allenby and Herkert, note 19 above, 167–82, at 167; as well as Marchant, note 19 above, at 19.

[28] See also G. Van Calster, D. M. Bowman and J. D'Silva, 'Protecting consumers or failing them? The regulation of nanotechnologies in the EU' *European Journal of Consumer Law* 1(1) (2011), 85–113, at 110. As well as B. R. Allenby, 'Governance and technology systems: the challenge of emerging technologies' in: Marchant, Allenby and Herkert, ibid., 3–18, at 13, Moses, note 24 above, at 265, as well as L. B. Moses, 'Sui generis rules' in: Marchant, Allenby and Herkert, ibid., 77–94, at 77.

complex the technology, the more stakeholders are involved[29] in the regulatory process, which in the end leads to the slowing down and complication of the technology.[30] As has been pointed out, 'the gap between technology and regulation has continued to grow'.[31] This is because of procedural requirements; most often highly controversial and polarized debates on the use of a specific technology and uncertainties involved in a technology and increasing judicial review, also referred to as the 'regulate, litigate, regulate, litigate' syndrome.[32]

The underlying regulatory trend in the EU is a shift from a purely interest group analysis and regulation towards a more institutional design.[33] This has been the case particularly following the introduction of the Single European Act in 1986 and the increase in the Union's competence to regulate in the environmental area, which goes hand in hand with the growing preference for risk aversion among the public.[34] A 'regulatory mix', consisting of command and control regulation, economic instruments, self-regulation and voluntarism has replaced the 'traditional' regulatory approach of prescriptive rules and general principles.[35] In 2006, the EU Better Regulation initiative explicitly recognized the need 'to adopt to the fast pace of technological change' and initiated another move forward in this regard.[36] Criticism of the increase in regulatory action thereby often relates to regulation being a 'pretext for

---

[29] For a list of stakeholders frequently involved in the regulatory process, refer to R. Baldwin, M. Cave and M. Lodge, 'Regulation: the field and the developing agenda' in: R. Baldwin, M. Cave and M. Lodge (eds) *The Oxford Handbook of Regulation* (Oxford: Oxford University Press, 2010), 3–16, at 4.

[30] Marchant, note 19 above, at 23.

[31] Gaudet and Marchant, note 27 above, at 167.

[32] Ibid., 169.

[33] Baldwin, Cave and Lodge, note 29 above, at 11.

[34] G. Van Calster, 'Risk regulation, EU law and emerging technologies: smother or smooth?' *NanoEthics* 2(1) (2008), 61–71, at 62.

[35] Lierman and Veuchelen, note 24 above, at 99; as well as N. Gunningham and D. Sinclair, 'Regulatory pluralism: designing policy mixes for environmental protection' *Law & Policy* 21(1) (1999), 49–76, at 49f. Traditional command and control legislation (Directives and Regulations at the European level) however enjoy high credibility and legitimacy with the public. See also Van Calster, Bowman and D'Silva, note 28 above, at 110.

[36] European Commission, 'Better Regulation – simply explained' (Luxembourg: Office for Official Publications of the European Communities, 2006), at 1.

protectionism',[37] thus introducing barriers to trade and obstacles to economic growth.[38]

This chapter will not dive into the discussion on the definition of either the term 'technology' or 'regulation', but will employ a rather broad definition of both. Koops defines a technology as 'the wide range of tools and crafts that people use to change or adapt to their environment', and regulation as 'the intentional influencing of someone's or something's behavior'.[39] This broad definition[40] avoids 'arguments at the margin'.[41] The definition of 'technology' does not refer to the aspect of newness, one of the focuses of this chapter. Another definition of technology points to the 'introduction of novelty through human practice'.[42] The underlying hypothesis for this volume in this regard is that the degree of 'newness' of a technology relates to the scientific uncertainties involved.[43] The more scientific uncertainties are associated with a technology, the 'newer' the technology.[44] As Moses puts forward: '[a] study of newness would not be a study of the regulation of technology, ... but rather a

---

37 D. Beyleveld and R. Brownsword, 'Complex technology, complex calculations: uses and abuses of precautionary reasoning in law' in: P. Sollie and M. Duwell (eds) *Evaluating New Technologies: Methodological Problems for the Ethical Assessment of Technology Developments* (Dordrecht: Springer, 2009), 175–90, at 189.

38 Baldwin, Cave and Lodge, note 29 above, at 7.

39 B. J. Koops, 'Ten dimensions of technology regulation. Finding your bearings in the research space of an emerging discipline' in: M. Goodwin, B. J. Koops and R. Leenes (eds) *Dimensions of Technology Regulation* (Nijmegen: Wolf Legal Publishers, 2010), 309–24, at 311, as also cited in Moses, note 18 above, at 5.

40 For an overview of the range of meanings of technology and definitions, refer to Moses, note 3 above, at 591f.

41 S. Weatherill, 'The challenge of better regulation' in: S. Weatherill (ed.), *Better Regulation* (Oxford: Hart Publishing, 2007), 1–17, at 1, as also cited in G. Van Calster, 'An overview of regulatory innovation in the European Union' *Cambridge Yearbook of European Legal Studies* 11 (2009), 289–320, at 289. For a more detailed analysis of the word regulation and the associated limitations, please refer to G. A. Hodge, D. M. Bowman and A. D. Maynard, 'Introduction: the regulatory challenges for nanotechnologies' in: G. A. Hodge, D. M. Bowman and A. D. Maynard (eds) *International Handbook on Regulating Nanotechnologies* (Cheltenham: Edward Elgar Publishing, 2010), 3–24, at 12, as well as for a definition and demarcation of the key terms 'regulation', 'law', and 'governance': R. Brownsword, 'The age of regulatory governance and nanotechnologies' in: Hodge, Bowman and Maynard, ibid., 60–80, at 61–5.

42 Woyke, note 26 above, at 16.

43 See also ibid., 17.

44 See also Stokes, note 2 above, at 26.

study of law or regulation in the context of a new or changing technology'.[45] In order to determine what the 'new' in 'new technology' entails, it is necessary to consider the role of science within technologies.[46] Sound science is often referred to as the key determinant in order to establish whether a technology is 'safe' or 'risky'.[47]

We often turn to science in order to establish the 'truth',[48] and as Oreskes phrases it, 'to tell us if a problem is real, and if so what to do about it'.[49] But is there really such a thing as sound science? It has been noted in the context of science and GMOs:

> Where, as in Biotech Products, the science relating to the safety of GM crops is contested, how is the matter to be resolved? A blameless response is that the question should be determined by reference to the view supported by 'sound science', this being taken to be a neutral and reliable arbiter. However, for many commentators on the practice and politics of science, this is a naive view. Science simply does not function this way. Scientists reasonably disagree with one another, not just about the bottom-line questions, but about matters of methodology, relevance, focus, and so on.[50]

This is the case regarding the underlying scientific debate regarding shale gas. For example, there is general agreement about the fact that monitoring activities are needed. What is subject to scientific debate and scientific uncertainty is when, how and to what extent the monitoring should be carried out in order to be effective.[51]

---

[45] Moses, note 18 above, at 7.

[46] Especially since when analysing the newness of a technology, it often becomes apparent that the same type of problem has already arisen in another context or another technology, and is hence no longer, as such, 'new'. See also Moses, note 3 above, at 597.

[47] M. van Asselt, E. Vos and T. Fox, 'Regulating technologies and the uncertainty paradox' in: Goodwin, Koops and Leenes, note 39 above, 261–86, at 275; as well as Stokes, note 2 above, at 26.

[48] Oreskes, note 26 above, at 369.

[49] Ibid., 370.

[50] Beyleveld and Brownsword, note 37 above, at 187f.

[51] E.g., the scientific discussion surrounding the presentation on 'Science-based policy for shale gas exploitation in Canada', J. Cherry, 'Shale gas in a low-carbon Europe: the role of research' organized by the DG for Research and Innovation of the European Commission on 23.02.2015, Brussels.

Some scholars are of the opinion that science can never be theory neutral.[52] Even if one does not necessarily agree with this rigid position on the role of science,[53] in the end it is society and not science[54] which determines whether a technology will be publicly accepted and how it will be controlled.[55] The same conclusion in the field of nuclear power is drawn: 'technical rationales do not help society to choose a side'.[56] Regarding shale gas, the same conclusion applies; society is to a large extent more than critical of the technology, and negative media headlines reinforce the public rejection of the technology.[57] Science is thus only laying the basis for this decision-making process. The relation between science and law is a controversial one. Whereas science is future-bound and focusing on new or emerging (un)certainties, the law is in the past and at best present-bound.[58] It is codifying everything as it is and in that way focusing on past problems.[59] Legal certainty is thereby relative, as law is subject to political decision-making, as well as open to interpretation and room for manoeuvre. 'Legal

---

[52] Beyleveld and Brownsword, note 37 above, at 187f. Oreskes has a similar perception claiming that the idea that science can prove proofs is a misunderstanding: Oreskes, note 26 above, at 369.

[53] Everson and Vis for example are of the opinion that regulators tend to perceive and use science as an objective and neutral source for decision-making, and that not only in theory but also in practice, as for example regarding scientific advice provided by the European Food Safety Authority (EFSA). M. Everson and E. Vos, 'Scientification of politics or politicisation of science?' in: M. Everson and E. Vos (eds) *Uncertain Risks Regulated* (London: Routledge/Cavendish Publishing, 2009), at 1–17, as well as E. Vos, 'Legal accountability: risk regulation and judicial review' presented at the ACELG Workshop on Regulating risks in the European Union – The co-production of expert and executive power, Amsterdam, 21–22. 05.2015.

[54] Jonas, note 20 above, at 36.

[55] M. Tallacchini, 'Before and beyond the precautionary principle: epistemology of uncertainty in science and law' *Toxicology and Applied Pharmacology* 207(2) (2005), 645–51, at 650; a practical example in this regard is Monsanto and its attempts to grow new types of genetically modified crops in the EU. See for further explanation C. Dunmore, 'Monsanto to withdraw EU approval requests for new GMO crops', *Reuters,* 17.06.2013, available at www.reuters.com/article/2013/07/17/us-eu-monsanto-gmos-idUSBRE96G16R20130717; last accessed 23.06.2015.

[56] G. Meskens, 'The trouble with justification – Getting straight on the science and politics of nuclear energy' *Energy Strategy Reviews* 1(4) (2013), 233–42, at 234.

[57] See also section 5.4 below.

[58] See also Tranter, note 25 above, at 755.

[59] Tallacchini, note 55 above, at 650.

rules or political decisions simply represent the normative framework for scientific content autonomously established by scientists'.[60] In applying these theoretical considerations to shale gas, it becomes apparent that the role of science in this technology is indeed a problem. A multitude of scientific studies have been carried out in multiple jurisdictions such as the US, the EU and its Member States (for example Germany, Poland and the UK); from different disciplines such as engineering,[61] economics,[62] law[63] and social science;[64] from different organizations (private,[65] public,[66] research,[67] non-governmental,[68]

---

[60] Ibid., at 647. This further developed in a German case regarding the competences of the Parliament to regulate technologies which can have adverse impacts on the entironment. The BverfG in BVerfG, Beschluss vom 08.08.1978 – 2 BvL 8/77 (Kalkar – Beschluss) claims that the adoption of the technical development of a technology is left to the administrative and judicial branch. The court states (at 135) that:

> Durch die Verwendung unbestimmter Rechtsbegriffe werden die Schwierigkeiten der verbindlichen Konkretisierung und der laufenden Anpassung an die wissenschaftliche und technische Entwicklung mehr oder weniger auf die administrative und – so weit es zu Rechtsstreitigkeiten kommt – auf die judikative Ebene verlagert. Behörden und Gerichte müssen mithin das Regelungsdefizit der normativen Ebene ausgleichen.

The BverfG further develops in Kalkar II (BVerfGE 81, 310) that regulation should go as far as reasonable precautionary measures permit: 'Fragen des anlageninternen Notfallschutzes, … haben … eine Minimierung des Risikos jenseits der Schwelle dessen zum Ziel, wogegen nach den Maßstäben der praktischen Vernunft Vorsorge getroffen werden muß' (at 122).

[61] L. Gandossi, *An Overview of Hydraulic Fracturing and Other Formation Stimulation Technologies for Shale Gas Production* (JRC, 2013).

[62] P. Mathis, R. Hugman, H. Vidas, A. Ritchie, T. Phung and J. Kisielewicz, *Macroeconomic Impacts of Shale Gas Extraction in the EU* (IGF GHK, 2014).

[63] For example, M. Ballesteros, F. Pelsy and L. Reins, *Regulatory Provisions Governing Key Aspects of Unconventional Gas Extraction in Selected Member States* (Milieu Ltd., 2013).

[64] P. Upham, A. Lis, H. Riesch and P. Stankiewicz, 'Addressing social representations in socio-technical transitions with the case of shale gas' *Environmental Innovation and Societal Transitions* 16 (September 2015), 120–41.

[65] Such as Ballesteros, Pelsy and Reins, note 63 above.

[66] Such as Gandossi, note 61 above.

[67] For example, J. Broderick, K. Anderson, R. Wood, P. Gilbert and M. Sharmina, *Shale Gas: An Updated Assessment of Environmental and Climate Change Impacts* (Tyndall Centre, 2011).

[68] A. Krupnick, H. Gordon and S. Olmstead, *Pathways to Dialogue – What Experts Say about the Environmental Risks of Shale Gas Development* (Resources for the Future, 2013).

governmental[69] or a mix thereof), which were hence focusing on different aspects of the activity (financing, technological problems, environmental and health problems and regulatory problems). Not surprisingly, these studies come to different results, ranging from complete denial of any scientific uncertainties associated with the activity and problems of scientific evidence,[70] to claims that shale gas will at the very least destroy the local environment.[71] In line with this stands an underlying problem for the scientific debate surrounding shale gas: the tremendous differences in scientific quality of the contributions to the subject matter.[72] There are abundant amounts of 'propaganda', 'junk science', 'dishonorable nonsense', 'honorable nonsense' publications and 'scientific' studies on shale gas, while at the same time 'irreproducible', 'immature' and at best 'mature' science in the very field is rare.[73] Thus, indeed the question emerges whether and how these different scientific outcomes can be identified and balanced; as well as whether one should afford higher authority and importance to some of the scientific problems rather than to others? Is there really that much scientific uncertainty in shale gas, and how does this matter for its regulation? In addition, the lack of sound science within the shale gas technology debate (if that indeed were to be the case), as Van Calster concludes for nanotechnology, 'could be remedied relatively easily by taking the necessary risk management steps if and when more

---

[69] EPA, *Assessment of the Potential Impacts of Hydraulic Fracturing for Oil and Gas on Drinking Water Resources* (EPA, 2015).

[70] Cherry, note 51 above.

[71] Y. Omorogbe for example, concludes that shale gas fracking is 'an environmentally hazardous practice that is a violation of the rights to life, and to a healthy environment'. Y. Omorogbe, 'Energy and water: African strategies' in: A. McHarg, D. N. Zillman, A. Bradbrook and L. Barrera-Hernandez (eds) *The Law of Energy Underground: Understanding New Developments in Subsurface Production, Transmission, and Storage* (Oxford: Oxford University Press, 2014), 469–90, at 489.

[72] This is attributed to the phenomenon that there is a political dimension to the scientific debate. In these cases science is (mis)used as an argument for conflicting political and moral claims. See also Oreskes, note 26 above, at 380. The same holds true also for nanotechnology: see D. A. Kysar, 'It might have been: risk, precaution and opportunity costs' *Journal of Land Use & Environmental Law* 22(1) (2006), 1–57, at 2.

[73] Cherry, note 51 above.

scientifically robust data on risks becomes available'.[74] As such, '[i]n risk regulation there are two types of expertise closely intertwined to the point they are almost indistinguishable – the expertise in evaluating risk and the expertise in governing'.[75] Taking this suggestion into account, and in order to answer the outlined question, one needs to consider the role of law within science, more precisely the role of the prevention and precautionary principles in the shale gas regulatory debate, as well as the timing of regulatory intervention.

## 2.1 New Technology and its Regulation: When?

'Time' is not only an important aspect regarding the revision of existing legal regimes (see above).[76] Within the regulation of new technologies, the 'time' variable also becomes decisive in determining the time for regulatory intervention.[77] In social science this is referred to as the Collingridge dilemma relating to the control of new technologies. Accordingly:

> Attempting to control a technology is difficult, and not rarely impossible, because during its early stages, when it can be controlled, not enough can be known about its harmful social consequences to warrant controlling its

---

[74] G. Van Calster, D. Bowman and J. D'Silva, "Trust me, I'm a regulator': the (in)adequacy of EU legislative instruments for three nanotechnologies categories' in: Goodwin, Koops and Leenes, note 39 above, 207–38, at 226, as well as Van Calster, Bowman and D'Silva, note 28 above, at 105.

[75] E. Fisher, 'Expert executive power, administrative constitutionalism and co-production: why they matter', presented at the ACELG Workshop on Regulating risks in the European Union, note 2 above, at 8; see for an analysis of the difficulties arising out of the close symbiosis between experts and politicians: M. Lee, *EU Environmental Law, Governance and Decision-Making*, 2nd ed. (Oxford and Portland: Hart Publishing, 2014), at 43f.

[76] Other interactions of law and technologies are for example the historical, cultural and social dimensions and mediation. These are discussed in Tranter, note 25 above.

[77] See also: L. B. Moses, 'Agents of change: how the law "copes" with technological change' *Griffith Law Review* 20(4) (2011), 763–94 and Moses, note 3 above, at 597. As well as G. Majone, 'Political institutions and the principle of precaution' in: J. B. Wiener, M. D. Rogers, J. K. Hammitt and P. H. Sand (eds), *The Reality of Precaution: Comparing Approaches to Risk Regulation in the United States and Europe* (RFF Press, 2010), 411–33, at 420.

> development. By the time these consequences are apparent, control has become costly and slow.[78]

Translating this social science-originating dilemma into legal terms, the question to lawyers is essentially about the right point of regulatory intervention within a technological development. Regulatory intervention at a premature stage is difficult because of possibly 'insufficient, conflicting or confusing data about the nature and impact of the new technology'[79] and the legal agenda and implications behind it.[80] The adverse impact on a society or the environment is not clear, or rather subject to uncertainty. An efficient regulatory regime may thus not be possible to establish at this point. If however the regulatory intervention comes too late in the technology development chain, the technology will have become entrenched in the society and 'influence and change become correspondingly more difficult [slow and expensive[81]] to effect'.[82] Arguably from a legal perspective, the precautionary principle interferes in the first part of the dilemma. It provides a general framework for law and policymakers to intervene, even if there are still uncertainties associated with a technology, without meaning 'when in doubt opt out'.[83]

Applying this rather theoretical framework to shale gas regulation, one can conclude that the US has waited too long for regulatory intervention and control. It has pursued a laissez-faire, trial-and-error approach to shale gas regulation[84] and now faces opposition from both the general public, as well as industry representatives. This makes it increasingly difficult for regulatory authorities to introduce changes into the existing regime. Opposition comes from both 'parties'; industry claims regulatory intervention is too strict and costly to implement, whereas NGOs claim it

---

[78] D. Collingridge, *The Social Control of Technology* (New York: St. Martin's Press, 1980), at 19.

[79] Laurie, Harmon and Arzuaga, note 17 above, at 5f.

[80] Moses, note 3 above, at 595.

[81] Moses, note 18 above, at 8.

[82] Laurie, Harmon and Arzuaga, note 17 above, at 5f.

[83] Van Calster, Bowman and D'Silva, note 74 above, at 230. Sunstein is however of another opinion. He argues that the precautionary principle paralyses because risks are everywhere and the principle in itself forbids action. The application of the principle is hence only prioritizing and shifting risks. See C. R. Sunstein, *Laws of Fear – Beyond the Precautionary Principle* (Cambridge: Cambridge University Press, 2005), at 26f.

[84] See Chapter 2.6.

does not go far enough in terms of environmental protection. Litigation regarding shale gas cases is increasing.[85]

The EU has taken the opposite approach and seems to have learned from the US's trial and error. Interestingly, without explicit recourse to the precautionary principle (see further below), the Commission adopted a non-binding Recommendation to escape, to some extent, the Collingridge dilemma,[86] as well as proportionality, subsidiarity and related competence discussions.[87] It remains to be seen whether the regulatory intervention happened too early, especially considering that there is no practical experience yet regarding the commercial use of hydraulic fracturing activity in the Union. Only exploratory drillings with experimental 'fracking' have taken place to date.[88] The conclusion Van Calster has drawn regarding nanotechnology, also has some value in the shale gas regulatory debate: 'a theoretical examination of the regulatory frameworks and their applicability to nanotechnologies is not in itself enough to decide whether or not the frameworks are adequate for dealing with nanotechnologies'.[89] However, also in the case of nanotechnology, non-binding instruments (in this case even self-regulatory ones), have prevented an early implementation of binding regulation and were broadly welcomed by the legislature[90] and other stakeholders.[91]

---

85 Norton Rose Fulbright, 'Analysis of litigation involving shale and hydraulic fracturing', 01.06.2014, available at www.nortonrosefulbright.com/files/20140101-analysis-of-litigation-involving-shale-hydraulic-fracturing-104256.pdf; last accessed 22.10.2014. Van Calster, Bowman and D'Silva, note 74 above, at 230.

86 As a binding proposal could take years of negotiations: see C. G. Molyneux and L. Tosoni, 'What's in a name? Legal implications of the EU Recommendation on shale gas' *Environmental Law Institute* 44(5) (2014), 10351–3.

87 For a detailed discussion see Chapter 2.

88 For example in Poland, see: Lewino-1G2: San Leon Energy, *Successful Vertical Frac Leads to Horizontal Well*, 23 January 2014, available at www.sanleonenergy.com/media-centre/news-releases/2014/january/23/lewino-1g2-successful-vertical-frac-leads-to-horizontal-well.aspx; last accessed 22.10.2014.

89 Van Calster, Bowman and D'Silva, note 74 above, at 231.

90 C. Meili and M. Widmer, 'Voluntary measures in nanotechnology risk governance: the difficulty of holding the wolf by the ears' in: Hodge, Bowman and Maynard, note 41 above, 446–61, at 455.

91 G. Van Calster and D. M. Bowman 'A good foundation? Regulatory oversight of nanotechnologies using cosmetics as a case study' in: ibid., 268–90, at 277.

What might actually distinguish the hydraulic fracturing or shale gas technology from other 'new' technologies usually caught by this concept or dilemma, is indeed the point of departure of the first part of the Collingridge equation. Moses points out that:

> At an early stage, [normally] little is known about the prospects for the new technology, the harms it might cause or the forms it might take. Thus regulators face an 'uncertainty paradox', where they are forced to make decisions in the absence of reliable risk information or foreknowledge of technological developments.[92]

This is arguably not the case for shale gas any more. Thanks to the US experience with shale gas, as well as to the multitude of scientific studies (even if they come to different conclusions), we do already have a decent amount of knowledge on the technology, as discussed in the following section.

## 3. RISKS, UNCERTAINTIES AND SCIENCE: THE ROLE OF THE PREVENTION AND PRECAUTIONARY PRINCIPLES

Shale gas is often referred to as a 'new'[93] or 'emerging'[94] technology or resource, which is associated with many knowledge gaps and scientific

---

[92] Moses, note 18 above, at 8.

[93] IPCC, *Climate Change 2014: Mitigation of Climate Change. Contribution of Working Group III to the Fifth Assessment Report of the Intergovernmental Panel on Climate Change* (Cambridge: Cambridge University Press, 2014), at 527. See also Argonne National Laboratory, 'Hydraulic Fracturing and Shale Gas Production: Technology, Impacts, and Regulation', available at www.afdc.energy.gov/uploads/publication/anl_hydraulic_fracturing.pdf; last accessed 26.06.2015, at 7, 12 and 18, also described as a 'new combination of technologies' in D. N. Zillman, A. McHarg, L. Barrera-Hernandez and A. Bradbrook, 'Introduction' in: McHarg, Zillman, Bradbrook and Barrera-Hernandez, note 71 above, 1–18 at 12 and 9, and as 'new' technology in T. Daintith, 'The common law of underground energy resources in the United States' in: ibid., 37–58, at 49; and H. Mostert and H. Meyer van den Berg, 'Roman-Dutch law, custodianship, and the African subsurface: the South African and Namibian experiences' in: ibid., 75–99, at 92.

[94] For example, U.S. Energy Information Administration, 'Review of Emerging Resources', available at www.eia.gov/analysis/studies/usshalegas/pdf/usshaleplays.pdf; last accessed 04.11.2014.

uncertainties about its environmental impacts in particular. Hence, the technology is often perceived to be a high risk technology. The 18th Coalition Agreement of the Grand Coalition in Germany, for example, refers to shale gas as a technology with an 'enormous risk potential'.[95] Along the same lines, Mall considers shale gas hydraulic fracturing as a 'big experiment without any actual solid scientific parameters guiding the experiment'.[96] However, one can question whether the technology (and its impacts) are really new and subject to uncertainties, or whether it is rather the likelihood of the impacts which is in question. The view of the President of the German Federal Institute for Geosciences and Natural Resources (BGR), Prof. Dr. Hans-Joachim Kümpel, hints in this direction: 'Critics speaking of an uncontrollable, high-risk technology in connection with fracking are simply wrong from a scientific perspective.'[97]

The perception of the risk potential of the technology thus differs among stakeholders and not only in terms of the classical industry/NGO divide.[98] From an academic point of view, the role of science, the status of shale gas as a 'risk' or 'new' technology, and the overarching relationship and context with the European environmental principles, especially the prevention and precautionary principle needs to be examined. This section will first describe the status of the precautionary principle in EU law generally, and then elaborate on its status within the shale gas regulatory debate specifically.

---

95 Koalitionsvertrag zwischen CDU, CSU und SPD, *Deutschlands Zukunft gestalten*, available at www.cdu.de/sites/default/files/media/dokumente/koalitionsvertrag.pdf, last accessed 04.11.2014.

96 Mall in C. Mooney, 'The truth about fracking' *Scientific American* 305 (2011), 80–85.

97 German Federal Institute for Geosciences and Natural Resources (BGR), *Kopenhagener Erklärung: Bei Schiefergas-Bewertung in European Commission den Sachverstand der Geologischen Dienste einbeziehen*, Press Release, 25.09.2014, available at www.bgr.bund.de/DE/Gemeinsames/Oeffentlichkeitsarbeit/Pressemitteilungen/BGR/bgr-140925_kopenhagener_erklaerung.html?nn=1542132 last accessed 04.11.2014 (translation by the author).

98 G. N. Mandel, 'Regulating emerging technologies' *Law, Innovation and Technology* 1(1) (2009), 75–92.

### 3.1 The Precautionary and Prevention Principle in EU Law[99]

The precautionary principle has been an integral part and a substantive principle of European environmental law since its inclusion in the Treaty of Maastricht in what is now Article 191(2) TFEU. With its origins in the German '*Vorsorgeprinzip*',[100] the principle aims at achieving a high level of environmental protection within the Union.[101]

The precautionary principle has been used by the European Institutions in the past to justify measures particularly in the areas of health protection,[102] nanotechnology[103] and food law.[104] Lacking an explicit definition in the Treaty and secondary law,[105] the European Court has mostly shaped the principle. Important cases in this regard are the 1996

---

[99] The text is adopted from G. Van Calster and L. Reins, *EU Environmental Law* (Cheltenham: Edward Elgar Publishing, 2017). This chapter focuses only on the status and implications of the precautionary principle in European law. The status and implications of the precautionary principle in U.S. law are not necessarily the same, and actually differ greatly in detail in some ways. This starts with the risk perception and risk attitude in both jurisdictions. See for example: E. U. Weber and J. S. Ancker, 'Risk perceptions and risk attitudes in the United States and Europe' in: Wiener, Rogers, Hammitt and Sand, note 77 above, 480–91. The reasons why the risk perception and also the precautionary principle are understood and interpreted differently are different patterns in culture, politics, legal systems, trade protectionism and perceptions and crisis events in both jurisdictions. See: J. B. Wiener, 'The real pattern of precaution' in: Wiener et al., ibid., 519–65, at 534–41. For the history and development of the principle in U.S., European and international law, please refer to: J.B. Wiener, 'The rhetoric of precaution' in: ibid., 3–38, at 9ff.

[100] For a description of the origins and history of the precautionary principle please see L. Kraemer, 'The genesis of EC environmental principles' in: R. Macrory (ed.) *Principles of European Environmental Law* (European Commission Law Publishing, 2004), 29–47, at 38f.

[101] For a detailed discussion of the principle please refer to N. de Sadeleer, *Environmental Principles: From Political Slogans to Legal Rules* (Oxford: Oxford University Press, 2005) at 91–223.

[102] Ibid., at 119. One of the first cases on the precautionary principle addressed the prohibition of substances in foodstuffs and the prohibition of imports of the latter on grounds of public health. See case ECJ, C-227/82 *Van Bennekom* [1983] ECR I-3883, at 37ff.

[103] Van Calster, note 34 above.

[104] ECJ, Case C-180/96 *United Kingdom v Commission* [1998] EHR I-226.

[105] This is not only the case in the EU, but also in the US. See: G. E. Marchant et al., 'Big issues for small stuff: nanotechnology regulation and risk management' *Jurimetrics* 52(3) (2012), 243–77, at 274ff.

*BSE* case,[106] the *Bergaderm*[107] judgment, as well as *Pfizer*[108] and *Gowan.*[109]

Whereas the Court in the *BSE* case did not explicitly refer to the principle at the time in upholding an export ban on beef,[110] it did so in 2000 during the *Bergaderm* judgment[111] for the first time. Limitations on the principle have been introduced by the Court in *Pfizer*,[112] stating that:

> it follows from the Community Courts' interpretation of the precautionary principle that a preventive measure may be taken only if the risk, although the reality and extent thereof have not been 'fully' demonstrated by conclusive scientific evidence, appears nevertheless to be adequately backed up by the scientific data available at the time when the measure was taken.[113]

It therefore follows that '[t]he precautionary principle can ... apply only in situations in which there is a risk, notably to human health, which, although it is not founded on mere hypotheses that have not been scientifically confirmed, has not yet been fully demonstrated'.[114] The Court in *Pfizer* stressed that scientific experts 'although they have

---

[106] ECJ, Case C-180/96, note 104 above, at 99.

[107] ECJ, Case C-353/98P *Laboratoires pharmaceutiques Bergaderm a.o. v Commission* [2000] ECR 5291.

[108] ECJ, T-13/99 *Pfizer Animal Health v Council* [2002] ECR II-3305. For a detailed discussion of the case see G. Van Calster, 'Note under case T-70/99, Alpharma Inc. v. Council of the European Union (Alpharma), 11 September 2002 and case T-13/99, Pfizer Animal Health SA v. Council of the European Union (Pfizer)' *Review of European Community & International Environmental Law* 12(1) (2003), 109–11.

[109] ECJ, Case C-77/09 *Gowan Comércio Internacional e Serviços* [2010] ECR I-13533.

[110] The Court stated that, '[w]here there is uncertainty as to the existence or extent of risks to human health, the institutions may take protective measures without having to wait until the reality and seriousness of those risks become fully apparent', ECJ, Case C-180/96, note 104 above, at 99. Other cases where the ECJ did not explicitly rely on the principle as such are for example regarding preservatives in foodstuffs. Case 53/80 *Kaasfabriek Eyssen*, regarding vitamins added to foodstuffs. Case 174/82 *Sandoz*, regarding pesticide residues in foodstuffs. These cases prioritized a high level of human health protection over the dismantling of trade barriers within the (now) Union.

[111] ECJ, Case C-353/98, note 107 above.

[112] *Pfizer*, note 108 above. For a detailed discussion of the case see Van Calster, note 108 above.

[113] *Pfizer*, ibid., para. 144.

[114] Ibid., para. 146.

scientific legitimacy, have neither democratic legitimacy nor political responsibilities. Scientific legitimacy is not a sufficient basis for the exercise of public authority.'[115] The 'Commission's exercise of public authority is rendered legitimate however as it is accountable to the European Parliament, and thus does have political responsibilities and democratic legitimacy'.[116] This in the end opens up the possibility for the Commission to act against a scientific opinion of the expert committee, as the Commission is not bound by it.

In *Commission v France,* the ECJ further clarified the limits and scope of application of the precautionary principle. The ECJ states that:

> a correct application of the precautionary principle presupposes identification of the potentially negative consequences for health of the proposed use of the substance at issue first, and a comprehensive assessment of the risk to health based on the most reliable scientific data available and the most recent results of international research second. Where it proves to be impossible to determine with certainty the existence or extent of the alleged risk because of the insufficiency, inconclusiveness or imprecision of the results of studies conducted, but the likelihood of real harm to public health persists should the risk materialize, the precautionary principle justifies the adoption of restrictive measures, provided they are non-discriminatory and objective.[117]

The wide room for manoeuvre and interpretation, due to the lack of a definition of the principle, as well as the fact that the principle is a 'knock-out' criterion for both positive and negative measures for the European Institutions, is often cited as criticism of the application of the principle.[118] The underlying problems relate to questions on the level and type of evidence of a potential adverse effect needed,[119] as well as which risk level is still acceptable and how and if benefits and risk are to be

---

115 Ibid., para. 201.

116 Ibid.

117 ECJ, Case C-333/08 *Commission v France* [2010] ECR I-757, at 92 and 93.

118 See also E. Fisher, 'Precaution, precaution everywhere: developing a common understanding of the precautionary principle in the European Community' *Maastricht Journal of European and Comparative Law* 9(7) (2002), 7–28.

119 A. Stirling, 'Science, precaution, and the politics of technological risk' *Annals of the New York Academy of Sciences* 1128(1) (2008), 95–110, at 101, as well as note 77 above, at 412.

balanced in decision-making.[120] Some elements of risk management are thus not adequately addressed in the European policy context. That this criticism is justified has been confirmed by the *Gowan*[121] case. A Portuguese company challenged the Commission's reliance on the precautionary principle as a justification for the restrictions imposed on the use of fenarimol, despite the fact that previously a favourable assessment and a favourable draft proposal had been issued by the Commission. The Court found that no 'manifest error of assessment' was committed by the Commission and accepted the Commission's position.[122]

The precautionary principle in European law is not only implemented in its primary legislation, but also in various pieces of secondary legislation, such as in the Annex of the Industrial Pollution Prevention Control (IPPC) Directive (now recast as the Industrial Emissions Directive (IED)) with reference to BAT (best available techniques),[123] the Water Framework Directive, in the area of food safety, as well as in the REACH Regulation.

Additionally, the main guidance and interpretation document on the precautionary principle, a Commission Communication[124] from 2000, establishes conditions and limits for the principle's application.[125] Accordingly, an action under the precautionary principle is only justified if the following three conditions are met:[126]

- identification of potentially adverse effects;
- evaluation of the scientific data available;
- evaluation of the extent of scientific uncertainty.

---

120 See also G. E. Marchant, D. J. Sylvester and K. W. Abbott, 'Risk management principles for nanotechnology' *NanoEthics* 2(1) (2008), 43–60; as well as R. F. Wilson, 'Nanotechnology: the challenge of regulating known unknowns' *The Journal of Law, Medicine & Ethics* 34(4) (2006), 704–13, at 710.

121 ECJ, Case C-77/09, note 109 above.

122 For a detailed discussion of the case, see A. Alemanno, 'Case C-79/09, Gowan Comércio Internacional e Serviços Lda v. Ministero della Salute, Judgment of the Court of Justice (Second Chamber) of 22 December 2010' *Common Market Law Review* 48(4) (2011), 1329–48.

123 On the role of BAT for shale gas regulation, see section 5.2.3.

124 European Commission, 'The precautionary principle', Brussels, 2.2.2000, COM(2000) 1.

125 For a critical analysis pointing at 'serious omissions' of the Communication, refer to Majone, note 77 above, at 420f.

126 European Commission, note 124 above, at 13ff.

Once a measure is deemed necessary under the precautionary principle, the following criteria[127] have to be met, which limit and balance a precautionary action. They derive from general principles of risk management. Actions or measures have to be:

- *proportional* in order to achieve the appropriate level of protection, thus they must generally not aim at zero risk, even if this might be required in some cases. Less restrictive alternatives shall be considered;
- *non-discriminatory,* comparable situations shall not be handled differently, and situations that differ shall not be handled uniformly;
- *consistent* with measures adopted in similar circumstances or using similar approaches;
- *based on an examination of the benefits and costs of action or lack of action,* meaning that they have to be subject to an economic cost–benefit analysis, and that negative and positive consequences as well as the overall cost have to be taken into account on both a short- and long-term basis;
- *subject to examination of scientific developments,* taking into account the overall societal risks and reviewing new scientific findings.

Only if all the criteria are fulfilled, is an action considered to adhere to the precautionary principle.[128]

The precautionary principle is therefore a tool for risk assessment and management. However, in stating that risk management should occur after 'a comprehensive assessment of risk' or that a scientific evaluation should at least be carried out,[129] the principle separates risk management from scientific evaluation.[130] This is especially striking considering that

---

[127] Ibid., at 17ff.

[128] In addition, Stirling establishes 17 features for a precautionary decision-making process, especially regarding new technologies. See Stirling, note 119 above, at 104.

[129] European Commission, note 124 above, at 14.

[130] See Fisher, note 75 above, at 22, as well as R. von Schomberg, 'The precautionary principle and its normative challenges' in: E. Fisher, J. Jones and R. von Schomberg (eds) *Applying the Precautionary Principle: Perspectives and Prospects* (Cheltenham: Edward Elgar Publishing, 2006), 19–42. See also on the role of science in the precautionary principle: P. Grandjean, 'Science for precautionary decision-making' in: EEA *Late Lessons from Early Warnings: Science, Precaution, Innovation* (European Environment Agency, 2013), 623–42.

scientific uncertainties and risk management are usually closely interrelated.[131] Indeed, the examples show that in cases where scientific uncertainties remain after the assessment of an activity, the precautionary principle opens the door for risk managers and policy-makers to take precautionary measures to reduce potential harm to health and the environment. Its main points of criticism,[132] it being potentially misused as a form of trade protectionism among them,[133] relate to the fact that the principle is sometimes misused in order to shift the burden of proof to the proponent of a technology, requiring the latter to demonstrate and prove the safety of a technology.[134] Consequently, if the proponent cannot satisfy this relatively high burden of proof, the principle can lead to bans and moratoriums on an emerging technology, even if in the end the technology brings about more environmental and societal benefits.[135]

A second important risk management principle is the prevention principle, which is often confused with precaution.[136] It was introduced in 1987 through the Single European Act, thus pre-dating the precautionary principle. The precautionary principle applies to *potential* risks, i.e., risks which cannot be qualified or quantified as yet because of a lack of scientific data or knowledge. By contrast, the prevention principle is referred to in situations where the likelihood of a risk occurring is quantifiable; the time and extent of the impact or damage not being known, however. The prevention principle is also included in Article 191(2) TFEU. In contrast to the precautionary principle, the prevention principle is an undisputed principle of public international law.[137] Its status is established and more widely recognized than the principle of

---

[131] Fisher, note 75 above, at 22

[132] See for a list of shortcomings: Majone, note 77 above, at 430.

[133] R. A. Edgell and R. Vogl, 'A theory of innovation: benefit, harm, and legal regimes' *Law, Innovation and Technology* 5(1) (2013), 21–53, at 35.

[134] Marchant, Sylvester and Abbott, note 120 above, at 46. Examples of the 'false' use of the principle are for example documented in: S. F. Hansen and J. A. Tickner, 'The precautionary principle and false alarms – lessons learned' in: EEA, note 130 above, 17–45.

[135] Nanotechnology is an example in this regard, see also Marchant, Sylvester and Abbott, note 120 above, at 46.

[136] See also Lierman and Veuchelen, note 24 above, at 99.

[137] See for a discussion of the relationship of the principles in international law: A. Trouwborst, 'Prevention, precaution, logic and law: the relationship between the precautionary principle and the preventative principle in international law and associated questions' *Erasmus Law Review* 2(2) (2009), 105–27.

precaution. It is an application of *sic utere tuo ut alienum non laedas*, as recognized for example in the *Trail Smelter* arbitrage.[138] At the European level, before being included in the 1987 Single European Act, the principle was already apparent in various Environmental Action Programmes, starting with the first one in 1972. It is thus a long-established and often applied principle of European environmental law.[139] As outlined above, the principle does not address potential risks on the grounds of scientific uncertainty, but risks which are known and likely to occur while carrying out a certain activity (or as a result of inaction).[140] However, similar to the precautionary principle, the application of the prevention principle is subject to two main thresholds, namely the actual relationship between the probability and the extent of damage, and a cost-benefit analysis.[141] The former aims at ensuring that the risk is considered with the required due diligence and ensures that a balance is struck with regard to the level and intensity of intervention and prevention.[142] The latter ensures that the risk is balanced with the economic effects of regulatory intervention (or the lack thereof), in the mid and long term.[143]

Even more than the precautionary principle, the prevention principle has its origins in waste and pollution legislation. The recast of the IPPC Directive, the EID Directive, is the main tool implementing the principle in secondary legislation. 'Prevention' is the overall topic of the Directive, as the explicit reference to the principle[144] in Recital 2 of the Preamble suggests. The second sentence of Article 1 states clearly:

> It [the Directive] also lays down rules designed to prevent or, where that is not practicable, to reduce emissions into air, water and land and to prevent the generation of waste, in order to achieve a high level of protection of the environment taken as a whole.

---

138 *US v Canada*, 3 RIAA (1941) 1907.

139 See also Kraemer, note 100 above, at 38; for a detailed discussion de Sadeleer as well, note 101 above, at 61–90.

140 See also on the distinction between precaution and prevention Kraemer, note 100 above, at 39. See also Van Calster, note 34 above at 66.

141 De Sadeleer, note 101 above, at 80f.

142 Ibid.

143 See above under the precautionary principle.

144 As well as the title of the old IPPC Directive.

This is also the main objective of the 2005 Thematic Strategy on the Prevention and Recycling of Waste,[145] which is currently under review.[146]

While the concept of prevention is straightforward in theory, the regulatory and political practice teach another lesson. That prevention is not crystal clear at all and subject to proportionality, political agreement and goodwill rather than science, has – once more – been demonstrated by the ECJ's ruling on the EU's Air Quality Directive.[147] The Directive[148] regulates the toxic gas nitrogen dioxide, which is the cause of around 29,000 deaths in the UK annually.[149] Applying a 100 per cent preventive approach would mean banning cars within city centres. Such a response however, would not appear appropriate and proportionate. Instead, the European Institutions, well aware of the risk, regulate this subject matter in the form of a Directive establishing limit values (thresholds) for air pollution in order to 'avoid, prevent or reduce harmful effects on human health and the environment as a whole' (Article 1 of the Directive).[150] More concretely, Article 13 of the Directive requires the Member States to keep nitrogen dioxide concentrations below certain limit values. In the case presented, the UK did not comply with this provision for some of its zones and agglomerations, nor did it apply for a postponement of five years according to Article 22 of the Directive. It did however draw up air quality plans as required under Article 23 of the Directive, and subsequently argued that it had *de facto* complied with the requirements of Article 13. The ECJ did not agree with this reasoning and held that:

> in order to be able to postpone by a maximum of five years the deadline specified by the directive for achieving conformity with the limit values for

---

[145] European Commission, 'Taking sustainable use of resources forward: A Thematic Strategy on the prevention and recycling of waste', 21.12.2005, COM(2005) 666.

[146] European Commission, 'Report on the Thematic Strategy on the Prevention and Recycling of Waste', 19.1.2011, COM(2011) 13.

[147] ECJ, C-404/13 *ClientEarth* nyr.

[148] Directive 2008/50/EC of 21 May 2008 on ambient air quality and cleaner air for Europe, [2008] OJ L 152/1.

[149] See ClientEarth, 'EU Court rules UK government must act to clean up deadly air pollution', Press Release, London, 19.11.2014, available at www.clientearth.org/news/press-releases/eu-court-rules-uk-government-must-act-to-clean-up-deadly-air-pollution-2699; last accessed 30.06.2015.

[150] As do many other Directives as well, for example the WFD on good environmental status and the Groundwater Directive establishing quality standards.

> nitrogen dioxide ... a Member State is required to make an application for postponement and to establish an air quality plan when it is objectively apparent, having regard to existing data, and notwithstanding the implementation by that Member State of appropriate pollution abatement measures, that conformity with those values cannot be achieved in a given zone or agglomeration by the specified deadline,

and further states that:

> it is for the national court having jurisdiction, should a case be brought before it, to take, ... any necessary measure, ... so that the authority establishes the plan required by the directive in accordance with the conditions laid down by the latter.[151]

That means that it is ultimately up to the national courts (not the Commission) to judge the suitability of the plans. The UK Supreme Court ordered the Secretary of State for Environment, Food and Rural Affairs to produce a plan to comply with the European legislation by the end of 2015.[152] From a prevention point of view, the judgment is arguably good news. It illustrates that ultimately it comes down to the application of science, rather than (political) negotiations in terms of compliance with legislation.

## 3.2 The Precautionary and Prevention Principles and their Roles within the Shale Gas Regulatory Debate

The role of law within science, as Harmon suggests, is the decision 'from a theoretical perspective, whether the law is an enabler or a prohibitor of technologies'.[153] At best, the law should promote the development of a technology without compromising the protection of the society and environment from uncertainties and risks of the associated technology.[154] As such, law and risk management at best 'neutralize the threat' or 'turn

---

151 ECJ, C-404/13, note 147 above.

152 UK Supreme Court, *R (on the application of ClientEarth) v Secretary of State for the Environment, Food and Rural Affairs* [2015] UKSC 28, on appeal from: [2012] EWCA Civ 897.

153 Laurie, Harmon and Arzuaga, note 17 above, at 10. See also: Moses, note 77 above, at 767.

154 Laurie, Harmon and Arzuaga, ibid., at 12 and 14f, as well as Stirling, note 119 above, at 96.

it to our advantage'.[155] The principles of precaution and prevention are the tools in European law to help strike this balance and implement this approach in practice. This section starts from the perspective that law in general, and more specifically at a Union level, imposes control on the management of 'known' risks via the invocation of the prevention principle; whereas regarding 'uncertainties' or 'unknowns', the law is more process rather than outcome-based and guided by the precautionary principle. As van Asselt concludes in a general manner:

> The [European] institutions use the fact that the experts are unable to provide this demand for certainty to label the problem as 'uncertain'. Subsequently, this labelling forms the basis for conducting a certain policy strategy, namely invoking the precautionary principle in a risk-aversive manner.[156]

Given controversial debates surrounding the technology and uncertainties of shale gas in science, politics and law (and indeed among the general public), it is surprising that the Commission Recommendation does not include any reference to 'precaution' or 'prevention'.[157] This omission is remarkable, considering the fact that the Preamble of the Recommendation refers to several general principles and rights, such as the 'right to life and the right to the integrity of the person, the freedom of expression and information, the right to conduct a business, the right to property, and [indeed] the high-level of health and environmental protection' (Recital 12 Preamble).

The Communication, which provides background information for the Recommendation, includes three references to 'precaution'. The term is twice used in describing the public concerns associated with shale gas: '[a]n important part of the population also perceives an insufficient level of precaution, transparency and public consultation in relation to shale

---

155 J. C. Pitt, 'Anticipating the unknown: the ethics of nanotechnology' in F. Jotterand (ed.) *Emerging Conceptual, Ethical and Policy Issues in Bionanotechnology* (Dordrecht: Springer, 2008), 103–16, at 113.

156 Van Asselt, Vos and Fox, note 47 above, at 277.

157 Furthermore, against this backdrop, it is even more surprising that Aldson is of the opinion that: 'the precautionary principle lie[s] at the heart of the Recommendation' and that this is 'evident throughout'. As reasoning, the Recommendation puts forward 'minimum principles' and that Member States are 'free to impose more stringent controls on fracking'. See F. Aldson, 'The EU's Recommendation on hydraulic fracturing and its implementation in the UK' *Environmental Liability* 2(3) (2014), 67–71, at 68. For the author of this volume, the opposite applies, and this is the result of a weak position of the principle. A strict application would have arguably resulted in a proposal of a Directive (however remember the competence issues discussed in Chapter 2).

gas activities',[158] and once regarding the fragmented application of existing legislation to shale gas activities by the Member States: '[t]he different approaches by public authorities may lead to a lack of level playing field and give rise to recurrent concerns regarding the suitability of the environmental safeguards and precaution measures'.[159] Thus, the lack of precaution and precautionary measures as such is recognized, however the 'answer' to these concerns does not mention the taking into account of the principle. The Commission thus merely recognizes the concerns, but does not take these up in the actual Recommendation.

Furthermore, the Commission Impact Assessment[160] contains several references to the principles, and seems to conflate them: '[i]n order to reassure the public and to follow a precautionary approach, preventive measures are more adapted and would be more effective if targeted at the sources of the problems, also entailing that the same measure could act on several problems'.[161] This underlines once more the struggle of how to adequately address the shale gas technology and the question of how much uncertainty there needs to be in order to take recourse to the precautionary principle. Additionally, in order to evoke the precautionary principle under normal conditions a risk assessment of the technology is needed.[162] In the case of shale gas an impact assessment, not a risk assessment, has been carried out. This might be another reason why the Commission refrained from referring to the principle. In order to rely upon the principle, uncertainty needs to be demonstrated. This might not be possible as there arguably is not so much uncertainty; hence recourse to the prevention principle is the safer option. As Van Calster (et al.) conclude and question regarding nanotechnology in several instances:

> Prima facie it is possible to assume based on the existence, structure and function of the regulatory regime that all cosmetics products incorporating engineered nanoparticles available in the European market are safe and do not

[158] European Commission, Communication on the exploration and production of hydrocarbons (such as shale gas) using high volume hydraulic fracturing in the EU, Brussels, 22.1.2014, COM(2014) 23 at 3 and only slightly different at 5.

[159] Ibid., 8.

[160] European Commission, 'Staff Working Document, Impact Assessment, Exploration and production of hydrocarbons (such as shale gas) using high volume, hydraulic fracturing in the EU', Brussels, 22.1.2014, SWD(2014)21.

[161] European Commission, note 160 above, at 41f.

[162] European Commission, note 124 above.

pose a risk to the health of the consumer when used for the purpose for which they were manufactured. But is it really that simple?[163]

The same reasoning applies to the case of shale gas. Drilling activities and even fracturing have taken place before. Hence we have an idea of the possible implications and an arguably solid legal framework is in place. However, as A Milieu Ltd. study on the regulatory regimes for shale gas in the EU Member States shows, there are certain gaps in the regulatory framework.[164] Still, for shale gas, compared to other technologies such as nanotechnology (further discussed below), the uncertainties are admittedly limited in number. For example, it is common knowledge that the injection of chemicals into the ground can have environmental impacts; just as underground wells can migrate, and deep and horizontal drilling can have impacts on the stability and the seismicity of rock formations. Shale gas has been drilled and fracked in a commercial manner in the US for close to ten years, hence a point of reference exists even if the experience cannot be directly applied to the European situation. It therefore seems that shale gas in the EU is not so much about the uncertainties involved, but rather about the likelihood that associated risks will emerge in practice.[165] This is a classical application of the confusion between 'policy-based evidence' (science has the function to underline existing commitments) and 'evidence-based policy' (the policy is based on science).[166] The approach of 'evidence-based policy' comes to the conclusion that a preventive policy approach is needed for shale gas, as the underlying science of the technology is rather straightforward.

---

[163] Van Calster, Bowman and D'Silva, note 74 above, at 213, and Van Calster and Bowman, note 91 above, at 272, and Van Calster, Bowman and D'Silva, note 28 above, at 91. This is similarly raised in O. Renn and A. Grobe, 'Risk governance in the field of nanotechnologies: core challenges of an integrative approach' in: Hodge, Bowman and Maynard, note 41 above, 484–507, at 502. In the following one can go further and ask the question of 'how safe is safe enough?' See also T. K. Epprecht, 'Producing safety or managing risks? How regulatory paradigms affect insurability' in: ibid., 163–76, at 164.

[164] Such as Ballesteros, Pelsy and Reins, note 63 above.

[165] See also Reins, note 87 above.

[166] See for a more detailed discussion of this phenomenon: J. Stilgoe, A. Irwin and K. Jones, *The Challenge Is to Embrace Different Forms of Expertise, to View Them as a Resource rather than a Burden – The Received Wisdom – Opening up Expert Advice* (London: Demos, 2006), at 17 and 23f.

Shale regulation may indeed not be caught by the precautionary principle, but rather by the prevention principle. The Commission seems to confuse, or interlink, 'precaution' and 'prevention'.[167] The continuing question mark over regulatory issues is whether scientific uncertainties necessarily require recourse to the precautionary principle directly. Is there really much uncertainty in shale? The overarching question thus is whether, at the very moment of a scientific uncertainty, one 'automatically' has to switch to the precautionary principle. With shale gas the uncertainty arguably is the degree to which one has to take measures, rather than the impacts as such. For example it is not in doubt that the injection of chemicals into the ground can have environmental impacts. It is rather the likelihood of impacts that are in question, not the possible impacts associated with the activity as such. Regarding cement deterioration of hydraulic fracturing wells for instance, the essence has been correctly summarized: '[a]ll wells can eventually leak. But under what circumstances will the leakage cause substantial harm?'[168]

This is not solely an academic discussion, our rather one which will also have considerable impacts on regulatory authorities and operators involved. This is especially relevant, if in a few years' time it turns out that the environmental impacts were underestimated and that the regulatory framework failed. If the scenario is one of precaution, then 'nobody knew with certainty' what could happen.[169] This arguably exonerates operators (unless of course they knowingly or negligently withheld information) should negative consequences occur eventually. For Member State authorities and the Commission, however, the outcome depends on the interpretation of the precautionary principle. There are arguably two ways of applying the principle in respect of new technologies. The common and predominant interpretation of the principle is the interpretation as 'an apology for regulatory intervention'. Precaution is employed to mitigate uncertainties and to mandate authorities to control the technology by taking recourse to the principle. This is how the principle has been exclusively applied in the Union to date. In this case there is no suggestion that an authority is bound by *duty* to apply the precautionary principle, and as a consequence could be sanctioned if the uncertain risks actually materialize. However, in another interpretation of the principle, this might indeed be the case. In a second, arguably less popular

[167] European Commission, note 160 above, at 41f.

[168] Cherry, note 51 above.

[169] For the underlying theory see also Jonas, note 20 above, at 37.

interpretation of the principle, is the precautionary principle functions as an 'obligation for action'. Authorities facing a scenario similar to that of shale gas can be held in breach of their obligation to act under the precautionary principle should it turn out that, even in a case of scientific uncertainty and a lack of knowledge, a certain threshold of scientific insight was exceeded and thus a limited amount of knowledge was available without the Commission acting upon it.

The outcome would be different if one relied on the prevention principle. With this principle, likelihood not scientific uncertainty is key. Authorities including the commission, have a duty of care vis-à-vis individuals, as also recently demonstrated in the *Urgenda* climate litigation.[170] They could potentially be held accountable for not having acted in the appropriate manner; that is for not having insisted on permit conditions which could have met the known risk of the permitted activity. Operators may consider themselves safe, as they arguably can rely on the permit defence.[171] The permit defence is '[an] event expressly authorised by, and fully in accordance with, the conditions of an authorisation given under applicable national laws and regulations'.[172] Its inclusion in the European Environmental Liability Directive was highly debated. Eventually it was left to the Member States to decide whether it should be applicable at all.[173] Not all (roughly two thirds) Member States transposed the defence into their national systems.[174]

---

[170] For a discussion of the case, refer to J. Lin, 'The first successful climate negligence case: a comment on Urgenda Foundation v. The State of the Netherlands', *Climate Law*, 2015 forthcoming. Available at http://papers.ssrn.com/sol3/papers.cfm?abstract_id=2626113; last accessed 09.07.2015.

[171] This would not apply of course should they withhold any information, in which case tort law would apply.

[172] European Commission, 'Environmental Liability Directive – Overview', available at http://ec.europa.eu/environment/legal/liability/pdf/eld_training/ELD%20Training%20Slides%20-%20Half%20Day.pdf; last accessed 02.04.2015, at 50.

[173] L. Bergkamp and A. van Bergeijk, 'Exceptions and defences' in: L. Bergkamp and B. Goldsmith, *The EU Environmental Liability Directive – A Commentary* (Oxford: Oxford University Press, 2013), 80–94, at 91

[174] Among the Member States that did not transpose the permit defence are Austria, Belgium (at the federal level), Bulgaria, Germany, France, Hungary, Ireland, Poland, Romania and Slovenia. See also Commission Report, 'Under Article 14(2) of Directive 2004/35/CE on the environmental liability with regard

The underlying understanding of the principle is that if the operator is given explicit and specific permission to carry out an activity, the law should be consistent and not impose liability for the consequences of a permitted activity.[175] The permit defence thus creates incentives for compliance and hence is a practical application of the prevention principle.[176] However, a valid point of critique of the principle is that an obtained permit is not a licence and excuse to cause environmental harm, and that it can be misused by operators not to reduce risks at the best level possible. Operators are incentivized to consider measures taken to be sufficient as long as the activity is in compliance with the permit conditions, even if operators could have gone further regarding the level of care and precaution.[177] Table 3.1 below illustrates these findings.

*Table 3.1 The precautionary and preventive principles and regulatory consequences for operators and authorities*

| Principle | | Operators | Authorities |
|---|---|---|---|
| Precaution | As 'apology for regulatory action' | No need to act | No need to act |
| | As 'obligation to act' | | Failure to fulfil obligations |
| Prevention | | Permit defence | Failure to fulfil obligations |

Arguably, in pursuing the second interpretation of the precautionary principle, the differentiation between both principles would further blur. Both principles would result in similar consequences, namely that action needs to be taken and that non-action can be sanctioned (as a failure to act). However in practical terms, when it comes to reasoning and establishing the failure to act of an authority, the threshold and burden of proof under the precautionary principle is higher and more difficult to establish, whereas for prevention this is more straightforward as risks and

to the prevention and remedying of environmental damage', Brussels, 12.10.2010, COM(2010) 581, at 4.

[175] Bergkamp, note 173 above, at 91. This however does not exclude instances of strict liability which a legislator included for the carrying out of certain activities, such as for example included in § 33 of the German Genetic Engineering Act (Gesetz zur Regelung der Gentechnik, as amended, 31 August 2015 (BGBl. I S. 1474)); or in § 22 of the Federal Water Act.

[176] Ibid.

[177] Ibid., at 92

not uncertainties are involved. The application of the precautionary principle as an 'obligation to take action' will most likely emerge at some point in the future, most probably after some more tragic event occurs due to a failure to apply the principle (e.g., people suffer food poisoning or a damage to health for example from controversial food, such as nanofood, because a ban has not been passed in time (unlike the BSE case, where a ban was enacted in time), or, indeed a shale gas facility blows up).

Operators have a comfortable position in both instances (precaution and prevention). Unless they withhold information, they can rest assured that the permit defence is available to them in a prevention scenario (which is exactly why the permit defence was so contested in the Environmental Liability Directive (ELD)) and all the more so in a precaution scenario.

What further adds to the complexity is that the European law- and policy-makers focus extensively on the US's experience with environmental impacts,[178] as there is comparatively little experience with exploration undertakings in the Union itself. Whereas this is certainly helpful to gain an overall impression of the activity, the lessons learned in the US cannot simply be carbon-copied and applied toward the European situation. Thus, in order to assess the real impacts, more exploratory drillings need to be carried out to further assess the exact implications, so as to act accordingly with regard to the precautionary and/or prevention principles. This is also what distinguishes shale gas from other – real – new technologies: the case of shale gas has already moved up the precautionary ladder; ten years ago would have been the time to discuss the scientific uncertainties involved, but due to, or thanks to, the US's trial and error, the European Institutions can fast-forward the discussion and directly move to the prevention stage, actual risk management.[179] The focus would then shift to proportionality discussions associated with the prevention principle.

---

[178] See alone in the reasoning and description of the environmental impacts on the Commission's Impact Assessment, note 160 above, at 18, 22, 25 and 37.

[179] For an overview of risk management challenges associated with shale gas refer to A. R. Lucas, T. Watson and E. Kimmel, 'Regulating multistage hydraulic fracturing: challenges in a mature oil and gas jurisdiction' in: McHarg, Zillman, Bradbrook and Barrera-Hernandez, note 71 above, 127–46, at 134ff.

# 4. CROSS-CUTTING 'NEW' TECHNOLOGIES AND THEIR REGULATION IN THE PAST

This section of the volume provides an introduction to the regulatory challenges associated with other cross-cutting 'new' technologies having a cross-title impact: CCS and nanotechnology. It has been claimed that usually States use a 'reoccurring pattern' of policy instruments for regulation in the environmental area.[180] Accordingly, 'States usually start out with a top-heavy, command and control approach, characterized in that early form at least by direct regulation.'[181] Due to regulatory failures such as economic inefficiency, environmental ineffectiveness and democratic illegitimacy,[182] the framework then shifts to a 'bottom-up approach to environmental action, focusing on specific actors rather than on regulatory mechanisms, and a shift to new instruments'.[183] Whether this paradigm holds true for the technologies of CCS, nanotechnology and shale gas, is examined in the next two sections.

## 4.1 The Regulatory Framework of the Carbon Capture Storage Technology – An Integrated Approach to Regulation?

Within the last two decades, several 'new' technologies[184] have emerged in the EU's energy and environmental area. Among these are the CCS technology,[185] which aims at, as its name suggests, capturing $CO_2$ from large point sources (power plants) and transporting it to a storage site to deposit or store the $CO_2$ in underground geological formations.[186] The main issues associated with the technology are carbon leakage into the

---

180 See Van Calster, note 41 above, at 290.

181 Ibid.

182 See ibid. also, for more references.

183 Ibid.

184 CCS is referred to as being 'relatively new': M. J. Mace, C. Hendriks and R. Coenraads, 'Regulatory challenges to the implementation of carbon capture and geological storage within the European Union under EU and international law' *International Journal of Greenhouse Gas Control* 1(2) (2007), 253–60, at 253.

185 Also described as a 'family of technologies': Zillman et al., note 93 above, at 13.

186 For a detailed description of the procedure and the individual elements of capture, transport and storage, please refer to C. Fouillac, '$CO_2$ capture, transport and storage, a promising technology for limiting climate change' in: J. B. Saulnier and M. D. Varella (eds) *Global Change, Energy Issues and Regulation Policies*, (Dordrecht: Springer, 2013), 121–41; as well as J. Gibbins and

atmosphere and the scientific uncertainties and knowledge gaps regarding long-term security of the geological storage sites.[187] In the EU, it is estimated that by 2030 the CCS technology could foster 15 per cent of the carbon emission reductions required.[188] The Commission purports that an integrated approach for the regulation of CCS exists.[189] Due to its technological and financial challenges, the technology involves high risks, including regulatory ones.[190]

The CCS Directive, a *sui generis* rule on the geological storage of carbon dioxide,[191] establishes a legal framework for the environmentally safe geological storage of carbon dioxide to contribute to the battle against climate change.[192] The purpose of environmentally safe geological storage is permanent containment of $CO_2$, to prevent and eliminate negative effects as well as any risks to the environment and human health.[193] Even if the technology differs to a large extent from the shale

---

H. Chalmers, 'Carbon capture and storage' *Energy Policy* 36(12) (2008) 4317–22.

[187] For a detailed explanation of the process, as well as a review of the environmental challenges and scientific uncertainties of this activity, see IPPC, 'Special Report on the Carbon Capture and Storage, Summary for Policymakers', available at www.ipcc-wg3.de/special-reports/.files-images/SRCCS-SummaryforPolicymakers.pdf; last accessed 22.10.2014, as well as W. Grevers and L. Luten, 'Introduction to the CCS chain: technological aspects and safety risks' in: M. M. Roggenkamp and E. Woerdman (eds) *Legal Design of Carbon Capture and Storage: Developments in the Netherlands from an International and EU Perspective* (Antwerp: Intersentia, 2009), 6–14.

[188] See Recital 5 of the CCS Directive.

[189] See European Commission, DG CLIMA, 'Ensuring safe and environmentally sound CCS', available at http://ec.europa.eu/clima/policies/lowcarbon/ccs/index_en.html last accessed 22.10.2014.

[190] A. McHarg and M. Poustie, 'Risk, regulation, and carbon capture and storage: the United Kingdom experience' in: McHarg, Zillman, Bradbrook and Barrera-Hernandez, note 71 above, 249–73, at 249.

[191] For a discussion on the benefits and criticism of *sui generis* rules, refer to Moses, note 28 above.

[192] For an analysis of the structure and content of the Directive and the individual provisions, refer to M. Doppelhammer, 'The CCS Directive, its implementation and the co-financing of CCS and RES demonstration projects under the emissions trading system (NER 300 Process)' in: I. Havercroft, R. Macrory and R. B. Stewart (eds) *Carbon Capture and Storage: Emerging Legal and Regulatory Issues* (London: Hart Publishing, 2011), 93–103, at 94–100.

[193] Article 1 CCS Directive, note 188 above.

gas extraction activities,[194] the CCS and shale gas technologies affect several elements of the environment, for example, air, water resources and geological formations. Furthermore, regarding both technologies, the determination and handling of risks and uncertainties is, and has been, an important topic of discussion within the regulatory and political process. Within CCS, *ex ante* determination of the risks is more complicated, as there are no past experiences to draw from, especially as to the long-term consequences involved (which is arguably also the case for shale gas). However, for CCS the European Commission itself promotes that 'although the components of CCS are all known and deployed at commercial scale, integrated systems are new, and a clear regulatory framework is required. The EU's CCS Directive provides such a clear regulatory framework.'[195]

The aim of this chapter is not to review and assess the regulatory framework applicable to CCS in an exhaustive manner,[196] but to see whether the framework provides parallels in regulatory approaches which are transferable to shale gas activities, especially in light of regulating scientific uncertainties associated with 'new' technologies. Hence, it starts from the presumption that the Commission's statement, while perhaps not 100 per cent accurate, contains some truth. Indeed, the Directive creates a clear and integrated legal framework for the activity. The existence of several guidance documents explaining the provisions of the Directive, among them '$CO_2$ Storage Life Cycle Risk Management Framework',[197] 'Characterisation of the Storage Complex, $CO_2$ Stream Composition, Monitoring and Corrective Measures',[198] 'Criteria for

---

[194] Haszeldine outlines similarities and differences of both technologies: S. Haszeldine, 'Geological factors in framing legislation to enable and regulate storage of carbon dioxide deep in the ground' in: I. Havercroft, R. Macrory and R. B. Stewart (eds) *Carbon Capture and Storage: Emerging Legal and Regulatory Issues* (London: Hart Publishing, 2011), 7–23, at 9f.

[195] See European Commission, note 189 above.

[196] This is for example done in the contributions in Roggenkamp and Woerdman, note 13 above.

[197] European Communities, 'Implementation of Directive 2009/31/EC on the Geological Storage of Carbon Dioxide Guidance Document 1 – $CO_2$ Storage Life Cycle Risk Management Framework', 2011, available at http://ec.europa.eu/clima/policies/lowcarbon/ccs/implementation/docs/gd1_en.pdf; last accessed 22.10.2014.

[198] European Communities, 'Implementation of Directive 2009/31/EC on the Geological Storage of Carbon Dioxide Guidance Document 2 – Characterisation of the Storage Complex, $CO_2$ Stream Composition, Monitoring and Corrective

Transfer of Responsibility to the Competent Authority',[199] 'Financial Security (Article 19) and Financial Mechanism (Article 20)',[200] underlines the assumption that for CCS activities such an integrated regime exists.[201] According to Article 18 of the Directive, the minimum period for the transfer of responsibility from the operator to the competent authority is 20 years.[202] A financial contribution has to be made available from the operator to the authority before the responsibility transfer which covers the monitoring costs for at least 30 years after the transfer (Article 20).

In addition, the Directive establishes a harmonized application procedure regarding permits and, consequently, harmonized conditions for the approval of a permit and the content thereof. So far Member States have different requirements in this regard. The fact that the Commission, as a superior authority, has some kind of say (even in the form of a non-binding opinion) in the issuance of such permits, pays tribute to the application of the precautionary principle and the (existing) scientific uncertainties. Nevertheless, it might also generate the erroneous impression of further safety, thereby contributing to greater public acceptance.[203] Industry questions this involvement, arguing that the Member

---

Measures', 2011, available at http://ec.europa.eu/clima/policies/lowcarbon/ccs/implementation/docs/gd2_en.pdf; last accessed 22.10.2014.

[199] European Communities, 'Implementation of Directive 2009/31/EC on the Geological Storage of Carbon Dioxide Guidance Document 3 – Criteria for Transfer of Responsibility to the Competent Authority', 2011, available at http://ec.europa.eu/clima/policies/lowcarbon/ccs/implementation/docs/gd3_en.pdf; last accessed 22.10.2014.

[200] European Communities, 'Implementation of Directive 2009/31/EC on the Geological Storage of Carbon Dioxide Guidance Document 4 – Article 19 Financial Security and Article 20 Financial Mechanism', 2011, available at http://ec.europa.eu/clima/policies/lowcarbon/ccs/implementation/docs/gd4_en.pdf; last accessed 22.10.2014.

[201] See also Havercroft and Macrory, note 11 above, at 300.

[202] For an analysis of the role of the competent authority in the regulation of CCS, refer to K. De Ridder and A. Haan, 'The role of the competent authority in regulating CCS' in: Roggenkamp and Woerdman, note 13 above, 301–19.

[203] Ricardo AEA, et al., 'Support to the review of Directive 2009/31/EC on the geological storage of carbon dioxide – Task 2: Assessment of the CCS Directive on the basis of evaluation and prospective questions – progress report', available at www.ccs-directive-evaluation.eu/assets/Summary-of-emerging-findings.pdf, last accessed 25.06.2015, at 13.

States' authorities are competent enough, and that Commission intervention would only slow down the procedure.[204]

The Directive further amends the Environmental Impact Assessment Directive;[205] the Water Framework Directive; the Directive on the limitation of emissions of certain pollutants into the air from large combustion plants;[206] the Environmental Liability Directive; the Waste Framework Directive; the IPPC (now IED) Directive and the Waste Shipment Regulation.[207] Thus, the Directive creates a specific regulatory regime for CCS activities by establishing an overall and general framework of the activity. In addition, it makes use of existing legal tools, such as the Environmental Impact Assessment and the Environmental Liability regime.

The clear contrast with shale gas is evident: in light of a lack of knowledge on the combined impact of components, the European Commission proposed a tailor-made regime for CCS, while the same lack of knowledge on the combined impacts has so far not led to binding regulatory action for shale gas whatsoever.

## 4.2 The Regulatory Framework for Nanotechnology – A Clear Case of Precaution?

Nanotechnology might not have a clear 'cross-title impact' at the regulatory level as such. However, it does at a practical application level, for example in the renewable and sustainable energy sector[208] (photovoltaic generation facilities and wind energy rotors), it also functions as

---

[204] Ibid., as well as Ricardo AEA, et al., 'Support to the review of Directive 2009/31/EC on the geological storage of carbon dioxide – Task 3: First Draft Recommendations', available at www.ccs-directive-evaluation.eu/assets/CCS-Directive-evaluation-Draft-recommendations.pdf; last accessed 25.06.2015, at. 8.

[205] Directive 2011/92/EU of 13 December 2011 on the assessment of the effects of certain public and private projects on the environment, [2012] OJ L 26/1.

[206] Directive 2001/80/EC of 23 October 2001 on the limitation of emissions of certain pollutants into the air from large combustion plants, [2011] OJ L 309/1.

[207] Regulation (EC) No. 1013/2006 of 14 June 2006 on shipments of waste, [2006] OJ L 190/1.

[208] Marchant et al., note 105 above, at 244.

an instrument for carbon capture in the fossil fuels industry.[209] In addition, it has been characterized as a technology with the potential to be 'profoundly transformative'.[210] It has also been stressed that nanotechnology should not 'becom[e] yet another future case study on how not to introduce a new technology'.[211] Nanotechnology regulation is exceptional, since, 'it is rare that so many novel legal issues are arrayed all at once'.[212] The case study of nanotechnology can be considered as a study of new technologies in general[213] and can be characterized as a 'trans-scientific problem'.[214] The case of nanotechnology offers three distinct angles to study new technologies.[215] First, the one 'as a social phenomenon, and as a way of framing meaning in public policy and regulatory debates'[216], second, the one 'as a set of multiple frontiers emerging from scientific disciplines'[217] and third, the one of regulation, 'where although the single word "regulation" is adopted, this word brings with it multiple meanings from different authors and in different contexts'.[218] Hence, the issues raised by nanotechnology are general questions associated with the regulation of new technologies:[219]

> Who is likely to bear the risks of any environmental challenges – investors, workers, or communities near the manufacturing plants? Who will reap the

---

[209] See also European Commission, 'Nano in Energy/Environment', available at europa.eu/research/industrial_technologies/nano-in-energy-environment_en.html, last accessed 25.06.2015.

[210] A. Nordmann, 'No future for nanotechnology? Historical development vs. global expansion' in: Jotterand (ed.), note 26 above, 43–63, at 43.

[211] S. F. Hansen, A. Maynard, A. Baun, J. A. Tickner and D. M. Bowman, 'Nanotechnology – early lessons from early warnings' in: EEA, note 130 above, 530–59, at 548.

[212] J. C. Monica, G. Van Calster and A. Patsa, 'A nanotechnology legal framework' in: M. Hull and D. Bowman (eds) *Nanotechnology Environmental Health and Safety – Risks, Regulation, and Management* (Elsevier Inc., 2014), 265–311.

[213] Royal Commission on Environmental Pollution, *Novel Materials in the Environment: The Case of Nanotechnology* (2008) at 7.

[214] Ibid., as well as A. Weinberg, 'Science and trans-science' *Minerva* 10(2) (1972), 209–22.

[215] Hodge, Bowman and Maynard, note 41 above at 5.

[216] Ibid.

[217] Ibid.

[218] Ibid.

[219] The question on the newness of the technology is however still apparent in scholarship. See for example ibid., 3.

> benefits …? How will decisions about risks and benefits be made, and by whom? What influences will shape those decisions?[220]

These are the same questions which are also debated regarding shale gas regulation. The complexity of shale gas is amplified by the discussion surrounding the horizontal and vertical competences, which for nanotechnology were not a prime concern.[221]

Indeed, nanotechnology serves as an interesting case study regarding the regulatory approach taken towards a new technology associated with scientific uncertainties.[222] At the beginning of the (regulatory) debate on nanotechnology, similar questions and problems emerged to the ones we are now facing in the shale gas regulatory debate.[223] More precisely, the first policy document on nanotechnology was, just as in the shale gas debate, a non-binding Recommendation on 'a Code of Conduct for responsible nanosciences and nanotechnologies research'.[224] Questions surrounding the adoption focused on the competences of the Union, the *de facto* and *de jure* effects of this policy measure as well as the content of the measure as such.[225] Similar to the debate within the European

---

[220] B. V. Lewenstein, 'What counts as a "social and ethical issue" in nanotechnology?' *HYLE – International Journal for Philosophy of Chemistry* 11(1) (2005), 5–18, at 9, as well as Stirling, note 119 above, at 98 and J. Wolff, 'Five types of risky situation' *Law, Innovation and Technology* 2(2) (2010), 151–63, at 154. For a discussion on the more ethical question of new technologies see: J. R. Herkert, 'Ethical challenges of emerging technologies' in: Marchant, Allenby and Herkert, note 19 above, 35–44, especially at 38ff, as well as Jonas, note 20 above, especially at 38f.

[221] See also further below.

[222] For an overview of the uncertainties associated with nanotechnologies, please refer to G. Van Calster and D. Bowman, 'Regulatory design for new technologies: Spaghetti Junction or Bauhaus Principles for regulating innovative cosmetic products' XXV *notizie di POLITEIA* 94 (2009), 75–93, at 79ff., as well as Q. Chadhury, H. Bouwmeester and R. F. Hertel, 'The current risk assessment paradigm in relation to the regulation of nanotechnologies' in: Hodge, Bowman and Maynard, note 41 above, 124–44, at 132f. For the different risk categories (occupational health and safety, production-related, consumer and environmental risks): T. Weidl, G. Klein, and R. Zöllner, 'The role of risk management frameworks and certification bodies' in: ibid., 462–83, at 464f.

[223] For an overview and discussion of the 'nano-product legal life cycle' and all regulatory issues involved, please refer to Monica, Van Calster and Patsa, note 212 above, at 266ff; as well as for an overview of the legal problems associated with nanotechnology, see Marchant et al., note 105 above, at 243.

[224] C(2008) 424, further assessed in detail below.

[225] See, e.g., I. Eisenberger, 'EU-Verhaltenskodex Nanotechnologie: Rechtsstaatliche und demokratische Aspekte' in: A. Gazsó and J. Haslinger (eds)

Institutions about the revision of the EIA Directive[226] and the inclusion of a mandatory EIA for shale gas, Parliament and Council could not reach an agreement on the EU Novel Foods Regulation[227] to include the modification of foods by new production processes, such as nanotechnology, under their scope.[228] In addition, also regarding nanotechnology, the regulatory approach taken is to a large extent based on 'conventional' regulatory experience,[229] as 'there has been limited action from regulatory decision-makers towards changing existing technology-neutral regulation to take the unique properties of these materials into account'.[230] This approach has also been taken regarding shale gas. In addition, in both case studies voluntary instruments have emerged as a reaction to public concerns. Voluntary reporting schemes on the import and production of nanomaterials emerged first in the UK, the US and Australia.[231] The practical effect of these schemes was rather limited,[232] so that at a later stage (Member) States such as France and California turned these into mandatory obligations.[233] Regarding shale gas, the development of the FracFocus Chemical Disclosure Registry in the US, managed by the Groundwater Protection Council and Interstate Oil and Gas Compact Commission of the United States, evolved from a voluntary industry initiative to an official means of chemical disclosure for

*Nano Risiko Governance* (Vienna: Springer, 2014), 165–82, at 11; as well as S. Vöneky and J. von Achenbach, 'Erste Stellungnahme zu der Empfehlung der Kommission für einen Verhaltenskodex für verantwortungsvolle Forschung im Bereich der Nanowissenschaften und -technologien' (Heidelberg: Max-Planck-Institut, 2008).

[226] Directive 2011/92/EU of 13 December 2011 on the assessment of the effects of certain public and private projects on the environment, [2012] OJ L 26/1.

[227] Regulation (EC) No. 258/97 of 27 January 1997 concerning novel foods and novel food ingredients, [1997] OJ L 43/1.

[228] Hansen et al., note 211 above, at 539.

[229] For an overview of the applicable 'media-specific environmental laws' to nanotechnology in the Union and a comparative assessment with the American, Japanese and Australian regulatory approaches, please refer to L. K. Breggin and J. Pendergrass, 'Regulation of nanoscale materials under media-specific environmental laws' in: Hodge, Bowman and Maynard, note 41 above, 342–71.

[230] Hansen et al., note 211 above, at 537.

[231] Ibid.

[232] Ibid.

[233] Ibid.

16 States.[234] Within the EU there is no comparable development yet, however.[235]

As for nanotechnology, there is no *overall* binding or non-binding regulatory tool, such as the Directive for CCS or the Recommendation for shale gas. On the contrary, the regulatory framework of nanotechnology is a mix of different (fragmented) non-binding and binding tools, applicable to the individual uses and substances, such as cosmetics or healthcare. This chapter will not address detailed case studies of the individual uses of nanotechnology[236] (e.g., chemicals, cosmetics,[237] food,[238] drug and medical devices,[239] military applications[240]) nor the different lenses and disciplines of law (business, intellectual property rights) as this has already been done elsewhere.[241] Instead, this chapter provides an analysis of the overall approach taken to the issue and establishes how the Commission has regulated this new technology in order to identify some lessons learned for the case of shale gas regulation.

---

234 See for more information FracFocus, available at http://fracfocus.org/; last accessed 26.06.2015.

235 See further section 5.1.2 on self-regulatory instruments.

236 This has been elsewhere, as for example in D. Bowman and G. Van Calster, 'Smoke that thunders – risk, confusion and regulatory frameworks for new technologies' in: L. K. Kjolberg and F. Wickson (eds) *Nano Goes Macro: Social Perspectives on Nanosciences and Technologies* (Singapore: Pan Stanford Publishing, 2010) at 359–86.

237 See for a discussion of the (nano-specific) regulatory framework on cosmetics for example Van Calster and Bowman, note 222 above, as well as Van Calster and Bowman, note 91 above, and Van Calster, Bowman and D'Silva, note 28 above, at 88–96.

238 See for a discussion of the (nano-specific) regulatory framework on foods for example, A. Gergely, Q. Chaudhry and D. M. Bowman, 'Regulatory perspectives on nanotechnologies in foods and food contact materials' in: Hodge, Bowman and Maynard, note 41 above, 321–41, as well as Van Calster, Bowman and D'Silva, note 28 above, at 96–100.

239 See for a discussion of the (nano-specific) regulatory framework on medical devices and medicine for example G. Van Calster and J. D'Silva, 'Taking temperature: a review of European Union regulation in nanomedicine' *European Journal of Health Law* 16(3) (2009), 249–69; as well as R. Sá Gaspar, 'Therapeutic products: regulating drugs and medicine' in: Hodge, Bowman and Maynard, note 41 above, 291–320, as well as Van Calster, Bowman and D'Silva, note 28 above, at 100–105.

240 See for a discussion of the (nano-specific) regulatory framework on military applications J. Altmann, 'Military applications: special conditions for regulation' in: Hodge, Bowman and Maynard, ibid., 372–88.

241 Hodge, Bowman and Maynard, ibid.

## 5. REGULATION OF 'NEW' TECHNOLOGIES: ALIGNING THE APPROACHES

Having briefly reviewed the applicable individual regulatory frameworks, this section will establish which lessons were, or can be, learned from the regulation of the CCS and nanotechnology with regard to the regulation of shale gas. Four dimensions will be examined:

1 the process and general direction of regulation;
2 risk management and the underlying principles;
3 the business case of the technologies; and
4 the public debate surrounding the individual technologies.

Table 3.2 below provides an overview of the stakeholders involved, the outcome of the regulatory process and the framework for the three comparative technologies.

*Table 3.2 Overview input and output of law and policy of CCS, shale gas and nanotechnology*

| Input | CCS | Nanotechnology[1] | Shale Gas |
|---|---|---|---|
| Stakeholders | CCS project developers, consultants, equipment manufacturers, governments and regulators, industry, NGOs, research centres and universities, storage plant operators, utilities | Research organizations, industrial organizations, public authorities, NGOs[2] | Companies, environmental and social NGOs, industry and trade associations, intergovernmental organizations, national authorities, regional or local authorities |
| Stakeholder meetings | – Four meetings of the European Climate Change Programme Working Group (ECCP) III on CCS in 2006<br>– European Technology Platform on Zero Emission Fossil Fuel Power Plants (ETP ZEP)<br>– Two meetings on review of the Directive (8.9. and 7.11. 2014)[3] | – A Preliminary Risk Analysis on the Basis of a Workshop Organized by the Health and Consumer Protection Directorate General of the European Commission[4] (1.3.2004)<br>– Technical workshop with stakeholder representatives[5] (30.6.2014)<br>– Sector specific initiatives such as annual reports of the EFSA Scientific Network of Risk Assessment of Nanotechnologies in Food and Feed[6] | European Science and Technology Network on Unconventional Hydrocarbon Extraction[7] Stakeholder meeting (7.6.2013, ongoing until 2017) |

*Table 3.2* (continued)

| Input | CCS | Nanotechnology[1] | Shale Gas |
|---|---|---|---|
| **Output** | | | |
| Implementation related | Implementation reports | Two implementation reports (2005–07 and 2007–09). | – Creation of shale gas-specific BREF<br>– Review of BREF on mining waste |
| Information related | Information Exchange Group | Four annual Nanotechnology 'Safety for Success' Dialogue Workshops<br>– Nanologue project (Facilitating the dialogue between research, business and civil society to improve the quality of life, create wealth and reduce impacts to society)<br>– Transparency measures Impact Assessment and public consultation<br>– Labelling of nanomaterials (sector-specific initiatives and CLP) | NGS Facts (similar to US FracFocus)[8] |
| Guidance documents | – Guidance Documents on $CO_2$ Storage Life Cycle Risk Management Framework<br>– Characterization of the Storage Complex, $CO_2$ Stream Composition, Monitoring and Corrective Measures<br>– Criteria for Transfer of Responsibility to the Competent Authority<br>– Financial Security (Art. 19) and Financial Mechanism (Art. 20) | – Sector-specific initiatives such as Guidance on the risk assessment of the application of nanoscience and nanotechnologies in the food and feed chain, or<br>– Guidance on the protection of the health and safety of workers from the potential risks related to nanomaterials at work and<br>– Working Safely with Manufactured Nano-materials – Guidance for Workers | – Guidance note on the application of Directive 85/337/EEC to projects related to the exploration and exploitation of unconventional hydrocarbons |

*Table 3.2* (continued)

| Input | CCS | Nanotechnology[1] | Shale Gas |
|---|---|---|---|
| **(Legislative) regulatory framework** | Directive 2009/31/EC | – Commission Recommendation 2011/696/EU on the definition of nanomaterial<br>– Commission Recommendation C(2008) 424 on a Code of Conduct for responsible nanosciences and nanotechnologies research<br>– REACH and CLP<br>– Sector-specific legislation, such as Regulation No. 1223/2009 on cosmetic products, Regulation No. 178/2002, Regulation No. 1333/200, Regulation 1169/2011 on food and Regulation No. 528/2012 on biocide products (BPR) | Commission Recommendation 2014/70/EU |

*Notes:*

1 For a complete overview of action, content and timeline of the individual regulatory instruments (Recommendation, CLP and REACH, as well as sector-specific instruments), please refer to Commission Communication, Second Regulatory Review on Nanomaterials, Brussels, 3.10.2012, COM(2012) 572, at 12–15. Table 3.2 provides just an indicative overview.

2 For a detailed overview of the NGOs involved in the regulatory process (and also on the European and Member States level): G. Miller and G. Scrinis, 'The role of NGOs in governing nanotechnologies: challenging the 'benefits versus risks' framing of nanotech innovation' in: Hodge, Bowman and Maynard, note 41 above, 409–45, at 420–7. As well as for a list of UK NGOs involved in the debate: A. Plows and M. Reinsborough, 'Nanobiotechnology and ethics: converging civil society discourses' in: Jotterand, note 26 above, 133–56, at 141.

3 For a list of participants of the stakeholder meeting, please refer to www.ccs-directive-evaluation.eu/documentation/; last accessed 25.06.2015.

4 For a list of workshop participants please refer to http://ec.europa.eu/health/ph_risk/documents/ev_20040301_en.pdf; last accessed 25.06.2015, at 7ff.

5 A report on the workshop is available at http://ec.europa.eu/enterprise/sectors/chemicals/reach/nanomaterials/index_en.htm#h2-7; last accessed 25.06.2015.

6 Available at www.efsa.europa.eu/en/topics/topic/nanotechnology.htm; last accessed 25.06.2015.

7 For a list of registered members available to the public, see www.foeeurope.org/sites/default/files/shale_gas/2015/carte_blanche_for_fracking_0.pdf; last accessed 25.06.2015.

8 Voluntary Hydraulic Fracturing Fluid and Additive Component Transparency Service, managed by the International Association of Oil and Gas Producers (IOGP). More information is available at www.ngsfacts.org/; last accessed 25.06.2015.

## 5.1 Process and General Direction of Regulation

### 5.1.1 Legislative history

The legislative history and context of the CCS technology evolved against the backdrop of CCS being perceived as an option to mitigate the impacts of climate change and to meet the 2°C global warming target.[242] Interestingly, the rationale for and impetus behind both technologies are comparable. The first parallel, the fact that shale gas is perceived by some scholars, scientists and experts as a tool to mitigate climate change,[243] might be subject to debate. The second parallel is not. Competitiveness of the economy and industry especially against the US[244] (as well as Australia in the case of CCS), is recognized as a key reason for the development of both technologies. The CCS Directive states that '[i]t would allow European industries to become leading players in a potentially burgeoning global market for CCS technology'.[245] Similar aspirations are held for shale gas.[246]

Considering the potential impact of the technology, the legislative history and context did generate little public reaction.[247] Indeed, compared to nanotechnology and shale gas, it went rather smoothly.[248] Several working groups were set up by the second and third European Climate Change Programmes (ECCP II and III), which stressed the need for a European CCS policy and legislative framework.

A separate framework was considered necessary, as the applicability of existing Directives, such as the Waste Framework Directive,[249] the

---

[242] European Commission, 'Staff Working Document, Accompanying document to the Proposal for a Directive on the geological storage of carbon dioxide – Impact Assessment', Brussels, 23.1.2008, COM(2008), at 2.

[243] For example ICF International, 'Mitigation of climate impacts of possible future shale gas extraction in the EU: available technologies, best practices and options for policy makers' (ICF International, 2014) available at http://ec.europa.eu/clima/policies/eccp/docs/mitigation_shale_gas_en.pdf; last accessed 30.06.2015.

[244] See for a comparison of law and policy approach to CCS in the US and Europe: E. J. Wilson and A. J. Gibbons, 'Deploying carbon capture and storage in Europe and the United States: a comparative analysis' *Journal for European Environmental & Planning Law* 4(5) (2007), 343–52.

[245] CCS Impact Assessment, note 250 above, at 15.

[246] Shale Gas Impact Assessment, note 160 above, at 15f.

[247] Further discussed below in section 5.4 on public concerns.

[248] For a detailed overview of the regulatory process and positions of the stakeholders involved, refer to Purdy and Havercroft, note 11 above, at 362ff.

[249] The Directive would potentially include captured $CO_2$ under the broad definition of waste, thus making it subject to specific waste permit requirements of Articles 9 and 10 of the Directive.

Landfill Directive[250] and the Water Framework Directive created some regulatory problems.[251]

As with many EU policy proposals, the process was accompanied by various stakeholder meetings and a public consultation on 'Capturing and storing $CO_2$ underground – should we be concerned?' in 2007. The consultation generated 787 replies and generally signalled support for the planned policy proposal to create an enabling framework for the activity.[252] Concerns mostly related to the fact that investments in CCS could delay or halt efforts in the areas of renewable energy and energy efficiency policy.[253] As such, most likely due to the public being unaware, the legislative history in the context of CCS does not include valuable lessons for the shale gas technology. For CCS, concerns were limited, as it did not really speak to the imagination of the public. The consultation process was not much different from the one for shale gas.[254] However, public awareness of the shale gas technology was high from the very beginning. This evidences once more that public awareness and concerns can have a key impact from the earliest stages of a technology's development.[255]

The start of the regulatory debates of nanotechnology and shale gas could not be more different. Regarding nanotechnology there was no discussion on whether this technology was needed or not. The debate focused more on the fact that the 'European excellence in nanosciences must finally be translated into commercially viable products and processes' and on the creation of a 'favourable environment for innovation' allowing the development in a 'safe and responsible manner'.[256] As such, regulatory intervention (on a European level) was not seen as a must. Ethical, health and safety as well as environmental concerns and risks, were to be studied, '*also* in order to *prepare* for possible regulation' (emphasis added).[257] Indeed, the 'regulatory process' of nanotechnology

---

250 The Directive bans all liquified waste at landfills.

251 See also Mace, Hendriks and Coenraads note 184 above, at 259.

252 CCS Impact Assessment, note 250 above, at 10.

253 Ibid., at 15.

254 See section 5.4 below.

255 See also P. Ashworth and C. Cormick, 'Enabling the social shaping of CCS technology' in: I. Havercroft, R. Macrory and R. B. Stewart (eds) *Carbon Capture and Storage: Emerging Legal and Regulatory Issues* (London: Hart Publishing, 2011), 251–63, at 254f. As well as section 5.4 for a more detailed discussion.

256 European Commission, 'Towards a European Strategy on Nanotechnology', COM(2004) 338, at 3.

257 Ibid.

moved relatively slowly, and arguably happened 'by accident'. The 2004 Strategy on nanotechnology identified five dynamics to stimulate nanotechnology progress, although none of these dynamics addressed or demanded the creation of a legal framework for the activity. Furthermore, the Nanotechnology Action Plan for Europe for 2005–09 did not include the need to review/establish a regulatory framework for the activity as an action area, and instead focused on research, development and innovation, as well as infrastructure and international cooperation (as identified in the 2004 strategy).[258] Later the Commission used a vociferatious approach. The first Implementation Report from 2007 on the Nanotechnology Action Plan provided an overview of the implementation of these measures.[259] It was only in 2008 that the regulatory aspects of nanotechnologies came to the centre of European attention, although the regulatory review started much earlier. The Commission documents on nanotechnology released in 2007 and 2008 took a long time to prepare and were withheld for some time. Finally, in February 2008 the Commission adopted the Recommendation on a Code of Conduct for responsible nanosciences and nanotechnologies research.[260] A couple of months later the Commission released the Communication on the regulatory aspects of nanomaterial.[261] The key conclusion of the review was that 'current legislation covers risks in relation to nanomaterials to a large extent and those risks can be dealt with under the current legislative framework'.[262] Adjustments regarding thresholds might be necessary in some cases, as well as strengthening the implementation of the legislation within the Member States' regulatory framework.[263] The European shale gas debate on the other hand was characterized by an immediate outcry for a ban on

[258] European Commission, 'Nanosciences and nanotechnologies: An action plan for Europe 2005–2009', COM(2005) 243, Brussels, 7.6.2005.

[259] European Commission, 'Implementation Report from 2007 on the nanotechnology action plan', Brussels, 6.9.2007, COM(2007) 505.

[260] European Commission, 'Recommendation on a code of conduct for responsible nanosciences and nanotechnologies research', Brussels, 07.02.2008, C(2008) 424.

[261] European Commission, 'Communication on the regulatory aspects of nanomaterials', Brussels, 17.6.2008, COM(2008) 366.

[262] Ibid., at 3.

[263] European Commission, note 269 above, at 9f.

the technology, largely although not exclusively by the general public.[264] A strict legislative regime was only asked for as a second option, mainly by more 'liberal' policy-makers and NGOs.[265]

### 5.1.2 Self-regulatory initiatives

Self-regulation is a rising phenomenon regarding the governance of new technologies.[266] Regarding nanotechnology, at least initially,[267] industry and NGOs had a cooperative attitude towards regulation and research.[268] For example, in the Union all stakeholders (including NGOs and industry representatives) were of the opinion that the existing legislation and databases addressing nanotechnology meet the objective of protecting confidential business information.[269] Moreover, in the United States, DuPont and Environmental Defense launched the Nano Risk Framework in 2007, which has served as a self-regulatory, best-practice framework for risk assessment on nanotechnology.[270] The framework contains six steps including risk assessment and review.[271] Other self-regulatory

---

264 For example local initiatives such as 'Gegen Gasbohren' in Germany available at www.gegen-gasbohren.de/ last accessed 26.06.2015, 'Fracturing Free Ireland' in Ireland http://frackingfreeireland.org/; last accessed 26.06.2015.

265 J. Potočnik, 'Transmission Note on the EU environmental legal framework applicable to shale gas', 26.01.2012, available at http://ec.europa.eu/environment/integration/energy/pdf/legal_assessment.pdf; last accessed 26.06.2015.

266 See for example for the different forms of self-regulation and motives: J. Sammeck, *A New Institutional Economics Perspective on Industry Self-Regulation* (Gabler Verlag, 2012), as well as N. Gunningham and J. Rees, 'Industry self-regulation: an institutional perspective' *Law & Policy* 19(4) (1997), 363–414, and L. Senden, 'Soft law, self-regulation and co-regulation in European law: where do they meet?' *Electronic Journal of Comparative Law* 9(1) (2005), 77–99. See also Lee, note 75 above, at 101–4.

267 The cooperative approach however, only lasted for a short time. In the end the classical NGO/industry approach dominated. As also further developed below and in Hodge, Bowman and Maynard, note 41 above at 4.

268 Mandel, note 98 above, at 7, as well as Van Calster and Bowman, note 91 above, at 277f.

269 RPA and BiPRO, *Summary of the public consultation on transparency for nanomaterials on the market* (2014), at 8.

270 For other self-regulatory approaches on nanotechnology, such as voluntary reporting and risk management schemes and systems, as well as codes of conduct, guidelines and auxiliaries, and an analysis thereof, please refer to Meili and Widmer, note 90 above, at 48.

271 Environmental Defense and DuPont, *Nano Risk Framework* (2007).

initiatives regarding nanotechnology have been developed by BASF, the Swiss Retailers Association and the Coalition of Non-Governmental Organizations, for example.[272]

Regardless of the initially cooperative approach, instead of the classical industry/NGO division and mistrust regarding regulation, the self-regulatory instruments have been criticized by some (mostly environmental) stakeholders because they 'usurp government oversight and public participation in the regulatory process'. [273] In the end, even if at first welcomed, including by authorities,[274] the approach failed and resulted in the opposite of effective self-regulation, namely the conclusion that self-regulation does not work at all.[275] The Code of Conduct on nanotechnology, further discussed below, might well serve as another example in the European Union in this regard. Unfortunately, also in the area of nanotechnology, '[d]espite a good start, it seems that ... research strategies are not leading to clear answers to critical questions; collaboration continues to be hampered by disciplinary and institutional barriers'.[276] This is certainly also the case for shale gas, which is characterized by the classical opposition of NGOs and industry, including on the issue of trade secrets and confidential business information.

Regarding shale gas, the industry has also reacted to concerns of the general public with a significant amount of self-regulation in the form of codes of conduct,[277] standards and best management practices, as well as

---

[272] See also Van Calster, Bowman and D'Silva, note 28 above, at 110, as well as for a detailed analysis of self-regulatory regimes on nanotechnology: D. M. Bowman and G. A. Hodge, 'Counting on codes: An examination of transnational codes as a regulatory governance mechanism for nanotechnologies' *Regulation & Governance* 3(2) (2009), 145–64, as well as more generally on voluntary programmes and partnerships on nanotechnology: Marchant et al., note 105 above, at 267f.

[273] Mandel, note 98 above, at 7, as well as J. Stilgoe, 'Governing nanotechnology in the EU: responsibility beyond risk' presented at the ACELG Workshop on *Regulating risks in the European Union – The co-production of expert and executive power*, Amsterdam, 21–22.05.2015 at 3.

[274] Meili and Widmer, note 90 above, at 455.

[275] As they mostly do not go beyond existing regulation. However, as they are swiftly implementable, they can serve as a 'bridge' towards binding legislation. Ibid., 458ff.

[276] Hansen et al., note 211 above, at 548.

[277] For a discussion of the different forms of codes of conduct as well as an analysis of its criticism, refer to B. Rappert, 'Pacing science and technology with codes of conduct: rethinking what works' in: Marchant, Allenby and Herkert, note 19 above, 109–26.

guidelines and principles, in both voluntary and mandatory forms.[278] Not all jurisdictions have self-regulatory guidance in place, but the more experienced States with the most technical knowledge and regulatory practice do. Table 3.3 provides an overview of self-regulatory practices and standards at the Member State level (Poland and UK), in the US and at international level.[279]

The majority of these instruments is drafted by industry associations, some of them (such as the CSSD guidance) in cooperation with authorities. The scope varies: whereas some instruments cover only the hydraulic fracturing activity as such, others go beyond this and also include best practices for community engagement and other activities related to the shale gas development process. This may very well (also) be an attempt to prevent the enactment of mandatory legislation, as industry obliged itself to follow certain rules and criteria and thus contributed to the regulatory regime.

### 5.1.3 Research and development context

The EU considers itself to be an 'Innovation Union',[280] pursuing the path towards a knowledge[281] and information society.[282] Between 2010–14 research funding in the Union saw an increase of 30 per cent, and after farming and regional development, was the third biggest budget line.[283] Whereas the Junker Commission might not uphold the entire cross-sectional research agenda which has led the European research path so

---

[278] See for a detailed discussion of these regulatory instruments in overall regulatory framework, especially in relation to command and control instruments: Gunningham and Sinclair, note 35 above, at 55ff.

[279] The analysis and comparison of the self-regulatory regimes would certainly be interesting and is a gap in research on shale gas as well, however due to the specific focus of this thesis, it is not covered in detail here.

[280] See also European Commission, 'Innovation Union', at http://ec.europa.eu/research/innovation-union/index_en.cfm?pg=intro; last accessed 25.06.2015, as well as critical on the Innovation Union and the European research policy: A. Stirling, 'Knowledge, evidence and power in EU technology governance from risk regulation through knowledge politics to innovation democracies' presented at the ACELG Workshop, note 2 above.

[281] European Commission, 'Towards a knowledge-based Europe', http://ec.europa.eu/information_society/newsroom/documents/catalogue_en.pdf; last accessed 25.06.2015.

[282] European Commission, Information Society, https://ec.europa.eu/jrc/en/science-area/information-society; last accessed 25.06.2015.

[283] R. L. Hudson et al., *The New European Commission: Bad News for Research and Innovation? An Analysis of the Juncker Commission* (Science Business Publishing Ltd., 2014), at 3.

*Table 3.3 Overview of self-regulatory practices and standards at the Member State, United States and international level*

| Level | Organization | Name | Year |
|---|---|---|---|
| Poland | Polish Exploration and Production Industry Organization (OPPPW) | OPPPW Oil & Gas Operating Recommended Practices | 2010 |
| United Kingdom | United Kingdom Onshore Operators Group (UKOOG) | Guidelines covering best practice for shale well operations in the UK | 2013 |
| | | Shale Community Engagement Charter | 2013 |
| United States | American Petroleum Institute (API) | Industry Guidance/Best Practices on Hydraulic Fracturing (HF), especially<br>• API HF1, Hydraulic Fracturing Operations-Well Construction and Integrity Guidelines<br>• API HF2, Water Management Associated with Hydraulic Fracturing<br>• API HF3, Practices for Mitigating Surface Impacts Associated with Hydraulic Fracturing | 2009–2011 |
| | Centre for Sustainable Shale Development (CSSD) | CSSD Performance Standards | 2013 |
| | Appalachian Shale Recommended Practices Group | Recommended Standards and Practices | 2012 |
| | Marcellus Shale Coalition | Recommended Practices | 2013 |
| | Investor Environmental Health Network (IEHN) | Shale gas exploration and production – Key issues and responsible business Practices – Guidance note for financiers | 2013 |
| | State Review of Oil and Natural Gas Environmental Regulations (STRONGER) | 2013 STRONGER Guidelines | 2013 |
| International | International Association of Oil & Gas Producers | Good Practice Guidelines for the development of shale oil and gas | 2013 |
| | DNV GL | Risk Management of Shale Gas Developments and Operations | 2013 |
| | International Risk Governance Council | Risk Governance Guidelines for Unconventional Gas Development | 2013 |
| | | Risk Governance Guidelines for Unconventional Gas Development (policy brief) | 2014 |

far, research in energy, digital technologies and pharmaceuticals is high on the agenda.[284] CCS, much like nanotechnology, has been subject to several research and funding programmes and initiatives. For example, the European Industrial Initiative (EII) on CCS was established as part of the Strategic Energy Technology (SET) Plan, together with the CESAR project[285] on the advancement of the technology as part of the Commissions FP 7 programme. The European Energy Programme for Recovery (EEPR) and the NER300[286] are programmes funded by European Emission Trading System (ETS) allowances.[287]

Regarding shale gas, research and development programmes on the activity were comprised mostly of studies commissioned by the European Institutions[288] and a European Science and Technology Network for unconventional hydrocarbons. The objectives of the network are:

> Structuring the dialogue among the stakeholders, fostering open information and knowledge sharing; presenting and discussing research activities and their results, as well as identifying gaps and R&D needs; examining knowledge gained from exploration and demonstration projects; identifying and assessing emerging technologies including their economic, environment and climate impacts.

In order to achieve these objectives 'all relevant stakeholders [are brought together] to foster a common understanding on relevant topics' and information is shared.[289] The network has been established for a three-year period from 2014–17 and will 'operate in full transparency. Accordingly, reports, technical documents, minutes of meetings and summaries of conferences, workshops and seminars will be made available on the internet.'[290] The network is divided into two overall working groups on

---

284 Ibid.

285 For more information see CESAR, '$Co_2$ Enhanced Separation and Recovery', available at www.co2cesar.eu; last accessed 25.06.2015.

286 For a detailed discussion of the programme refer to Doppelhammer, note 192 above, at 100ff.

287 European Commission, Communication on the Future of Carbon Capture and Storage in Europe, Brussels, 27.3.2013, COM(2013) 180 at 4.

288 See for an overview: European Commission, Energy and environment – Environmental Aspects on Unconventional Fossil Fuels – Related studies, available at http://ec.europa.eu/environment/integration/energy/uff_studies_en.htm; last accessed 05.07.2015.

289 European Science and Technology Network on Unconventional Gas, 'Mandate of 7 July 2014', available at https://ec.europa.eu/jrc/en/uh-network/mandate-rules; last accessed 05.07.2015.

290 Ibid.

'Exploration and demonstration and production projects in the EU' and 'Emerging technologies for well stimulation', which are further divided into task forces.[291] So far, the results of the task forces are rather limited; previous meetings centred around the general organization and working method. The working group meeting in June 2015 was more content-oriented and focused on the discussion of how to collect data on exploration and extraction activities in the Member States, particularly in Poland. It became apparent that one of the main concerns regarding data collection is related to business confidentiality and competition. The argument of industry representatives was that the research is still ongoing and quite new and that 'data is money'.[292] The Network on unconventional hydrocarbon extraction is supposed to bring together 'practitioners from industry, research, academia as well as civil society, so as to ensure a fair and balanced exchange of ideas'.[293] However it has been criticized as having been 'taken over by industry groups which are using it as a platform to promote fracking'.[294] Indeed, it became apparent that industry is (supposed to be) the key driver. This is due to industry possessing the data, but being reluctant to share it. The network is arguably stuck; the working group meetings in June did not foster the exchange of any

---

[291] Working group 1 is subdivided into the task forces: Collection of projects, Comparative assessment, Identifying gaps and needs in research, and the creation of an Annual Report; Working group 2 is subdivided into the task forces: Current technique, new technique, cost and PROs & CONs. The author is a member of task forces 3 and 4 of working group 1.

[292] European Science and Technology Network on Unconventional Hydrocarbon Extraction meeting, June 2015.

[293] European Commission, European Science and Technology Network on Unconventional Hydrocarbon Extraction, available at https://ec.europa.eu/jrc/en/uh-network, last accessed 25.06.2015, as well as the Mandate, note 298 above.

[294] A. Neslen, 'Green groups accuse EU shale gas panel of fracking lobby takeover' *The Guardian*, 15.04.2015, available at www.theguardian.com/environment/2015/apr/15/green-groups-accuse-eu-shale-gas-panel-of-fracking-lobby-takeover; last accessed 25.06.2015. According to Friends of the Earth, less than 10 per cent of the members are from civil society and more than 70 per cent 'either represent or have financial links to the fracking industry'. Friends of the Earth Europe, 'Carte blanche for fracking – How the European Commission's new advisory group is letting the shale gas industry set the agenda', *Corporate Europe Observator*, available at http://corporateeurope.org/sites/default/files/attachments/carte_blanche_for_fracking_final.pdf; last accessed 25.06.2015. This is however not completely correct, as the author is listed on the Friends of the Earth's list under her former affiliation at Milieu Ltd. which has been counted as belonging to 'industry' representatives. The author however, registered and introduced herself with KU Leuven, hence, her research affiliation.

useful (scientific) ideas. The strong representation of the industry is not confined to shale gas, but illustrates a systematic failure in European law-making.[295] One reason for this is the fact that industry has more financial capacity for 'lobbying' the institutions. NGOs and other civil society organizations often lack financial capacity[296] and frequently employ an ad hoc approach.[297]

Further consideration should be paid to the form and overall policy context of the regulatory intervention of nanotechnology. Even if both nanotechnology and shale gas were initially addressed through a non-binding Recommendation,[298] nanotechnology has been embedded in a broader policy context from the beginning. The subject matter has been connected to other policy areas relevant to the field, such as sustainability, area integration, education and research, technology transfer, international cooperation and public investment.[299] Shale gas is more isolated and typically treated only in relation to environmental protection and energy security policy. It has only slowly begun to drift into other areas, such as research and development. Shale gas has only recently been taken up in the Horizon 2020 programme,[300] and mostly within the scope of

---

295 See for example also in the case of synthetic biology, Stokes, note 3 above.

296 For the Commission-sponsored participation of one NGO in each working group. The NGOs were selected together with the EEB.

297 Also Van Calster, note 41 above, at 301, as well as Miller and Scrinis, note 243 above, at 436.

298 The regulatory framework of nanotechnology developed through an incremental approach including several sector-specific regulatory instruments.

299 G. Hunt, 'Nanotechnologies and society in Europe' in: G. Hunt and M. Mehta (eds) *Nanotechnology: Risk, Ethics and Law* (London: Earthscan, 2006), 92–104, at 93.

300 For example project 'Shale Environment – Maximizing the EU shale gas potential by minimizing its environmental footprint', Ref.: 640979 available at http://cordis.europa.eu/project/rcn/193771_en.html; last accessed 25.06.2015 and project 'M4ShaleGas – M4ShaleGas: Measuring, monitoring, mitigating, managing the environmental impact of shale gas', Ref.: 640715, available at http://cordis.europa.eu/project/rcn/193743_en.html; last accessed 25.06.2015 as well as project 'FracRisk – Furthering the Knowledge Base for Reducing the Environmental Footprint of Shale Gas Development (FracRisk)', Ref.: 636811, available at http://cordis.europa.eu/project/rcn/193422_en.html; last accessed 25.06.2015.

broader energy calls,[301] or in the environmental area.[302] It was part of FP7 only to a limited extent[303] and was not part of the European Strategic Energy Technology Plan (SET-Plan). On the contrary, research was one of the key priorities for nanotechnology. The NANOSAFE programme started under FP5 and was subject to several follow-up programmes (NANOSAFE2, NANO-Pathology and NANODERM),[304] and two projects relating to nanotechnology and its interaction with the environment and living world (STREP), as well as on human health and environmental toxic impacts (NANOTOX), both under FP6. This is certainly also because shale gas falls under the domain of big capital-intensive investments, whereas for nanotechnology, this is different. Due to low initial costs, start-ups and university spin-offs were able to engage in research and development activities quite easily.[305]

### 5.1.4 An incremental approach to regulation

One similarity in the regulatory approach taken regarding shale gas and nanotechnology is the idea of applying an incremental approach to

---

301 See among others, the calls for competitive low-carbon energy, available at http://ec.europa.eu/research/participants/portal/desktop/en/opportunities/h2020/calls/h2020-lce-2014-1.html; last accessed 25.06.2015.

302 See note 308 above.

303 For example project 'G-ETI – The Governance of Energy Technology Innovation: Investigating Regulatory Regimes in Shale Gas', Ref.: 331962, available at http://cordis.europa.eu/project/rcn/107123_en.html; last accessed 25.06.2015 and project 'HYDROFRAC – Computational Multiscale Modelling of Hydraulic Fracturing for Shale Gas Development', Ref.: 623667, available at http://cordis.europa.eu/project/rcn/185948_en.html; last accessed 25.06.2015 or in relation to CCS: for example project 'CO2SHALESTORE – CO2 Sorption and Flow in Shale Reservoirs', Ref.: 276862, available at http://cordis.europa.eu/project/rcn/98617_en.html; last accessed 25.06.2015 and project 'CO2SHALESTORE – CO2 storage in fractured shale', Record Number: 155858, available at http://cordis.europa.eu/result/rcn/155858_en.html; last accessed 25.06.2015.

304 See for more information also European Commission, 'Nanoscience and Nanotechnology in the Research Programmes of the European Community', available at ftp://ftp.cordis.europa.eu/pub/nanotechnology/docs/nanosciences_260702.pdf last accessed 25.06.2015.

305 See also section 5.3 below.

regulation for nanotechnology and shale gas. Whereas for nanotechnology, the creation[306] of the approach worked in practice,[307] this might not be so for shale gas.

As for nanotechnology, '[m]uch of the research performed on ultrafine particles in the 1990s now feeds into what we know about the potential risk of nanomaterials and lays the foundation for many of the current scientific research hypotheses in the field of nanotechnology'.[308] The same is true for shale gas: most of the knowledge is based on the experience gained with the extraction of conventional hydrocarbons. The same is true for the applicable legal frameworks. Several scientific studies and academics alike have come to the conclusion that nanotechnology is covered by the existing legal framework.[309] Initially, it was held that REACH 'sets the best possible framework for the risk management of nanomaterials when they occur as substances or mixtures',[310] under the premise that the regime would be modified to some extent.[311] Indeed, the emergence of the technologies has not resulted in the amendment and adoption of existing technology-neutral legislation,[312] integrating the specific characteristics and properties of the 'new' technologies into the existing baseline.[313] Only later was sector-specific legislation related to food,[314]

---

306 Not necessarily the practical application however.

307 For the key elements of this incremental approach on nanotechnology regulation, see Chadhury, Bouwmeester and Hertel, note 222 above, at 135ff.

308 Hansen et al., note 211 above, at 534.

309 Ibid., at 54 also for further references. At least at the beginning of the scientific debate.

310 European Commission, 'Nanomaterials', available at http://ec.europa.eu/enterprise/sectors/chemicals/reach/nanomaterials/index_en.htm; last accessed 25.06.2015

311 Ibid.

312 'Technology-neutral' regulation is indeed often seen as an efficient and frail framework for new technologies. See: Moses, note 24 above, at 239 and 70ff.

313 Hansen et al., note 211 above, at 541.

314 Regulation (EU) No. 1169/2011 of 25 October 2011 on the provision of food information to consumers, amending Regulations (EC) No. 1924/2006 and (EC) No. 1925/2006 and repealing Commission Directive 87/250/EEC, Council Directive 90/496/EEC, Commission Directive 1999/10/EC, Directive 2000/13/EC of the European Parliament and of the Council, Commission Directives 2002/67/EC and 2008/5/EC and Commission Regulation (EC) No. 608/2004, [2011] OJ L 304/18. See also G. Van Calster, 'Simply swallow? The application of nanotechnologies in European food law' *European Food and Feed Law Review* 4(30 (2009), 167–71.

cosmetics[315] and medical devices[316] adopted or modified to also include nanotechnology. The EU legislature decided in both cases that the existing legal framework is comprehensive enough to deal with new challenges, at least for now, and that the creation of a completely new regulatory regime is unnecessary.[317] This is the case, even if for both technologies public concerns about possible adverse impacts on human health and the environment exist.[318] Regarding nanotechnology, the concerns of NGOs relate to consumer protection requirements not being met, especially regarding labelling of products containing nanotechnology.[319] In the case of shale gas, concerns mostly relate to adverse environmental impacts such as the pollution of water and seismic events.[320]

The main pieces of legislation applicable to nanomaterial, aside from the sector-specific legislation relating to cosmetics, medical devices, food, etc., are the REACH and CLP Regulations,[321] which both encompass nanomaterial under the definition of 'substance'. The Commission

---

315 Regulation No. 1223/2009 on cosmetic products, [2009] OJ L 342/59. See also D. M. Bowman and G. Van Calster, 'Flawless or fallible? A review of the applicability of the European Union's cosmetics directive in relation to nano-cosmetics' *Stud. Ethics Law Technology* 2(30) (2008), 1–35, as well as D. M. Bowman, G. Van Calster and S. Friedrichs, 'Nanomaterials and regulation of cosmetics' *Nat. Nanotechnology* 5(2) (2010), 92, at 92.

316 Directive 2007/47/EC of 5 September 2007 amending Council Directive 90/385/EEC on the approximation of the laws of the Member States relating to active implantable medical devices, Council Directive 93/42/EEC concerning medical devices and Directive 98/8/EC concerning the placement of biocidal products on the market, [2007] OJ L 247/21. See also Van Calster and D'Silva, note 239 above.

317 See for nanotechnology: Van Calster, Bowman and D'Silva, note 74 above, at 225. Moses however, is of a different opinion, stating that nanotechnology poses unique problems, and that the existing regulatory regimes including pre-nanotechnology did not work effectively. See Moses, note 18 above, at 16.

318 Hansen et al., note 211 above, at 541.

319 RPA and BiPRO, note 277 above, at 9.

320 See Chapter 1.2.

321 See also Brussels, 3.10.2012, COM (2012) 572 at 5f, as well as for an assessment of the regulation of nanomaterials under REACH: Chadhury, Bouwmeester and Hertel, note 222 above, at 125f. and M. Widmer and C. Meili, 'Approaching the Nanoregulation Problem in Chemicals Legislation in the EU and US' in: Hodge, Bowman and Maynard, note 41 above, 238–67, at 243ff.

further states that it 'assess[es] possible further legislative modifications'.[322] Other applicable legislation in the environmental area (besides the sector-specific tools on e.g., cosmetics, etc.) includes further legislation on waste (e.g., the Waste Framework Directive, the Landfill Directive, the Sewage Sludge Directive, as well as the WEEE and RoHS Directives) and water (the Water Framework Directive, the Groundwater Directive and the Drinking and Urban Waste Water Directives). Additionally, the Seveso II Directive[323] and the Ecolabel Regulation include provisions applicable to nanomaterials as well. The law and policy framework on nanotechnology is thus a mix of sector-specific binding Regulations and Directives and several non-binding Recommendations on nanotechnology. Van Calster et al. refer to this incremental approach as 'nano-hooks', which have been attached to existing legislation, mostly after review of the different legislative acts.[324] This approach of creating hooks[325] in existing legislation has the advantage that, in contrast to the slow and oftentimes contested establishment process of an entirely new legislative instrument, it is supposed to be easier and more time-efficient.[326] This approach is almost certainly wishful thinking regarding shale gas. The first attempt of adopting a 'shale gas-hook' into existing legislation, more concretely the introduction of a mandatory EIA for shale gas activities in the EIA Directive, failed. It might take some time until policy-makers accept this idea as a potential regulatory tool for the activity.

*5.1.4.1 The role of the Recommendations within the incremental approach* A valuable lesson for shale gas might be learned by assessing how the two Recommendations on nanotechnology were implemented, as these are the only two exclusive regulatory instruments *directly* on nanotechnology. The instruments are the 2008 Code of Conduct on Nanotechnology and the 2011 Recommendation on the definition of

---

322 As of 02.02.2015, European Commission, Nanomaterials, available at http://ec.europa.eu/enterprise/sectors/chemicals/reach/nanomaterials/index_en.htm #h2-3; last accessed 02.02.2015.

323 Directive 96/82/EC of 9 December 1996 on the control of major-accident hazards involving dangerous substances, [2007] OJ L 10/3, as amended.

324 Van Calster, Bowman and D'Silva, note 74 above, at 226, as well as Van Calster and Bowman, note 222 above, at 75.

325 Also referred to as 'recombinant regulation', defined by Stokes as 'the recombination of existing legislation with new techno-scientific objects'. See Stokes, note 3 above.

326 Van Calster, Bowman and D'Silva, note 74 above, at 226; as well as Van Calster, Bowman and D'Silva, note 28 above, at 106.

nanomaterials. The Code of Conduct focuses on nanotechnology and risk assessment, precaution, research sustainability and competitiveness; thereby providing the general basis for an enabling framework for research and sound regulation. Quite atypically for a Recommendation, it also addresses researchers and research funding bodies,[327] in addition to the usual audience of the Member States (and the Commission itself). The addressees and the scope are thus broader those of the Commission Recommendation on shale gas, which solely focuses on environmental aspects and is addressed solely to the Member States. The Code of Conduct contains seven overall principles for a research framework on nanotechnology: public well-being, sustainability, precaution, democracy, excellence, innovation, and responsibility.[328] The description of the principles as such is vague; the precise content is hard to determine and the principles are partially overlapping and hard to differentiate.[329] For example, section 3.1 of the Code outlines that nanotechnology and nanomaterials (N&N) research activities should be 'conducted in the interest of the well-being of individuals and society in their design, implementation, dissemination and use'. Whereas this principle might read as commonly agreeable in theory, it is certainly not easy to apply in practice. The interests of individuals and society as a whole, if definable at all, might change over time and are certainly dependent on their context. In addition, Section 4.1.13 identifies a key priority in that 'funding bodies and organisations should encourage fields of N&N [nanotechnology and nanomaterials] research with the broadest possible positive impact'. Again, even if priorities are identified (environmental, consumer and worker benefits); it is far from clear what exactly a positive impact includes, and perspectives certainly differ on this. One can state that the Code of Conduct is an early example of principle-based regulation, without being called as such.[330] [331] From this consideration,

---

327 See for example the provisions of section 4.1.11–4.1.13 of the Code.

328 Section 3 of the Code.

329 Eisenberger, note 225 above.

330 For a discussion of principle based-regulation, the advantages and disadvantages and its application to new technologies, refer to R. B. Carter and G. E. Marchant, 'Principles-based regulation and emerging technology' in: Marchant, Allenby and Herkert, note 19 above, 157–66.

331 C. Meili and E. Mantovani, 'MasterPlan – Issues and Options on the Path Forward with the European Commission Code of Conduct on Responsible N&N Research' (NANOCODE, 2011), at 9.; as well as European Commission, 'Recommendation to the Member States on Responsible Research and Innovation', 14.09.2011, at 2, available at http://ec.europa.eu/smart-regulation/impact/

one could expect that the Code of Conduct is a successful example of a first regulatory approach to a new technology. However, the Code has never been revised, nor have Member States reported on its implementation and activities. The Netherlands is one of the few examples of Member States which have included the adherence to the EU Code of Conduct as one of the requirements for obtaining government funding.[332] A revision is not expected; the code will remain part of European nano-policy as it stands.[333]

The second nano-specific regulatory instrument is the Recommendation containing a definition of nanomaterials.[334] Nanomaterial means 'a natural, incidental or manufactured material containing particles, in an unbound state or as an aggregate or as an agglomerate and where, for 50 per cent or more of the particles in the number size distribution, one or more external dimensions is in the size range 1 nm–100 nm'.[335] Interestingly, the definition[336] was established in the form of a non-binding Recommendation in 2011, more than seven years after the very first policy step, the EuroNanoForum in 2003,[337] and the first official policy instrument in 2004; the Communication 'Towards a European

planned_ia/docs/2010_rtd_012_responsible_research_innovation_en.pdf, last accessed 12.06.2015 as well as Eisenberger, note 225 above.

332 Nanocode, 'Country Report The Netherlands', available at www.nanopodium.nl/CieMDN/content/NanoCode_Country_Report_NL_Final.pdf; last accessed 12.06.2015 at 23.

333 As well as Eisenberger, note 225 above.

334 See also D. M. Bowman, J. D'Silva and G. Van Calster, 'Defining nanomaterials for the purpose of regulation within the European Union' *The European Journal of Risk Regulation* 1(2) (2010), 115–22.

335 Commission Recommendation 2011/696/EU of 18 October 2011 on the definition of nanomaterial, [2011] OJ L 275/38. The definition is currently under review.

336 On the importance of definitions as a basis for regulating new technologies in general and regarding nanotechnology in particular, see D. Williams, 'The scientifc basis for regulating nanotechnologies' in: Hodge, Bowman and Maynard, note 41 above, 107–24, at 109–15.

337 The Forum brought the different stakeholders together for the very first time. It was sponsored by the EU Institutions, but they did not play an official role. See for more information: Hunt, note 307 above, at 93.

Strategy for Nanotechnology'.[338] The 2008 Communication on the regulatory aspects of nanotechnology still recognized the 'absence of generally accepted definitions [of] the term nanomaterials'.[339] The 2009 Parliament Resolution on regulatory aspects of nanomaterials[340] acknowledged the 'disagreement starting at the level of definitions' and consequently called for a 'comprehensive science-based definition of nanomaterials in Community legislation' but also at the international level.[341] The definition is currently subject to review.[342]

The definition provided the basis for defining nanomaterials as contained in the Biocides Regulation,[343] and envisaged being implemented in the regulatory framework on medical devices.[344] As part of the update to the novel foods legislation, the Parliament proposed amending the definition of 'novel food' to also include 'engineered nanomaterials' and by extension some aspects of the recommended definition.[345] The review

---

338 European Commission, 'Towards a European Strategy on Nanotechnology', COM(2004) 338.

339 European Commission, note 269 above, at 3. And indeed, it proved to be rather difficult to agree upon a common definition: Bowman, D'Silva and Van Calster, note 342 above, at 115. See also Marchant et al., note 105 above, at 260f.

340 European Parliament, 'Resolution of 24 April 2009 on regulatory aspects of nanomaterials' P6_TA(2009)0328.

341 Ibid., at F, 7 and 8.

342 See also JRC, 'Towards a review of the EC Recommendation for a definition of the term "nanomaterial", Part 1 on "Compilation of information concerning the experience with the definition"', available at https://ec.europa.eu/jrc/sites/default/files/lbna26567enn.pdf; last accessed 25.06.2015 and JRC, 'Towards a review of the EC Recommendation for a definition of the term "nanomaterial", Part 2 on "Assessment of collected information concerning the experience with the definition"', available at https://ec.europa.eu/jrc/sites/default/files/jrc_nm-def_report2_eur26744.pdf; last accessed 25.06.2015.

343 Regulation (EU) No. 528/2012 of 22 May 2012 concerning the making available on the market and use of biocidal products, [2012] OJ L 167/1.

344 European Commission, 'Proposal for a Regulation on medical devices, and amending Directive 2001/83/EC, Regulation (EC) No. 178/2002 and Regulation (EC) No. 1223/2009', Brussels, 26.9.2012, COM(2012) 542, as well as JRC Part 1, note 350 above, at 24.

345 More precisely, the Parliament suggested including the text:

> any intentionally produced material that has one or more dimensions of the order of 100 nm or less or is composed of discrete functional parts, either internally or at the surface, many of which have one or more dimensions of

of the Regulation however failed in 2011, and consequently, the existing definition was not amended.[346] The definition has not led to an overall common playing field per the definition of nanotechnology at member European level, nor at Member States' level.[347] In the REACH framework for example, the definition is used as a working definition, but has not been included in the Regulation, only in its guidance documents.[348] On a national level, there is no harmonized application of the definition.[349] Whereas Denmark implemented the complete definition,[350] France for example, in its Decree No. 2012-232 on the mandatory reporting of nanomaterials, adopted a definition which is 'mainly based on the EC Recommendation'. However, it restricts the '"substances with nanoscale status" only to substances intentionally manufactured at the nanoscale'.[351] Norway, which is not a Member State of the EU but of the EEA, departs from the definition suggested in the Recommendation and provides that 'only intentionally added nanomaterials need to be registered'.[352]

The *de facto* effect and experience with these two Recommendations allow for two conclusions to be drawn. First, the fact that the non-binding definition on nanotechnology has made its way into binding (sector-specific) European legislation with a direct effect on the Member States clearly illustrates the case for some *de facto* as well as *de jure* effects

---

> the order of 100 nm or less, including structures, agglomerates or aggregates, which may have a size above the order of 100 nm but retain properties that are characteristic to the nanoscale. Properties that are characteristic to the nanoscale include (i) those related to the large specific surface area of the materials considered and/or (ii) specific physico-chemical properties that are different from those of the non-nanoform of the same material.

As cited in JRC, 'Considerations on a Definition of Nanomaterial for Regulatory Purposes', (2010), at 15. See also for a more detailed discussion: Bowman, D'Silva and Van Calster, note 342 above, at 116ff.

346 See also Monica, Van Calster and Patsa, note 212 above, at 298.

347 See also ibid., 284.

348 European Commission, Nanomaterials in REACH and CLP, available at http://ec.europa.eu/enterprise/sectors/chemicals/reach/nanomaterials/index_en.htm; last accessed 25.06.2015.

349 One might argue at first glance that this is not really a problem as European law supersedes, however as the definition is established in the form of a non-binding Recommendation, this is not the case.

350 JRC note 350 above, at 22f.

351 Ibid.

352 Ibid., at 31. This report also contains further examples of how the Recommendation is implemented in the European Union but also beyond.

of the Recommendation. It is an example of how the content was made binding through other pieces of legislation. It is doubtful that a more-or-less uniform definition of nanotechnology (adopted in specific sectors such as cosmetics, biocides, etc.) could have been employed in European legislation without the Recommendation as an underlying basis.[353] The document cleared the way for legislation in this regard, despite transposition and implementation in the Member States having been rather limited. A similar effect might be expected to occur regarding shale gas.[354] Due to the problems of competences as established in Chapter 3, which essentially prevent the adoption of a 'Shale Gas Directive (or Regulation)', it might be an option to include individual provisions applicable to shale gas in existing legislation through amendments of the latter. Admittedly, the competence issues remain, but Member States such as the UK and Poland might be more likely to support (read: 'not veto') some regulation within the shale gas value chain in lieu of an entire legislative package. Furthermore, when only regulating some individual steps, the 'significance' threshold might not be a hurdle and thus the Qualified Majority Voting (QMV) decision-making might be applicable, rendering the passing of individual measures possible. The Commission might indeed be aiming for this, as the issues addressed in the Recommendation are closely related to the legal gaps identified by the Milieu Ltd. Report on regulatory gaps in the Member States.[355]

Secondly, the experience with the Nanotechnology Code of Conduct shows that the establishment of a principle-based regulatory regime in the form of a non-binding Recommendation failed.[356] The code had little to no impact; it was not even subjected to the announced review. Member

---

[353] For an overview of the different definitions in the European context alone, please refer to Bowman, D'Silva and Van Calster, note 342 above, at 121ff. Another example of a Recommendation which has become binding is Commission Recommendation 2003/361 on the definition of micro, small, and medium sized enterprises, OJ [2003] L 124/36. The definition is now included in Article 3 of the REACH Regulation. See also Molyneux and Tosoni, note 86 above.

[354] See also Molyneux and Tosoni, note 86 above.

[355] Ibid.

[356] See also Stilgoe, note 281 above, at 9. For a discussion on the different forms of principle-based regulation and the (dis)advantages please refer to:

States did not transpose and implement it at national level as its provisions are not specific enough. However, the code did embed nanotechnology in a broader policy context (research, sustainable development, innovation and accountability). This was probably also the main reason for considering the promotion of the code to an overall code for new technologies. In the end this was never implemented.

The Recommendation on shale gas, although labelled as establishing 'minimum principles', contains substantially more detailed provisions, and does not just establish overall principles like the code on nanotechnology. It is only focused on one subject matter (the environment) and has only one addressee (the Member States). It is hence more limited in application. On first thought one might conclude that the Recommendation is thus 'easier' for the Member States to apply in practice and, as the reactions from the reporting obligations of the Member States show, most Member States did implement the Recommendation in their national systems. This is the case even if at second glance it becomes apparent that the Recommendation contains many black letters without suggesting how the Member States should carry out the different requirements. For example, on how to determine or monitor the baseline for the individual elements (water and air quality, soil conditions, biodiversity, etc.).

The comparison of the two instruments shows that the level of precision of a Recommendation is relevant for its practical impact. As one scholar summarizes regarding nanotechnology, '[i]f there is a lesson for the regulation of nanotechnology from what has passed and is passing as "better regulation", it is the need to stress the substantive aspects of regulation'.[357] The Code of Conduct was an attempt (without being named as such) to introduce a principle-based regulatory approach. However it did not 'fly' as the content was not specific enough and had too many addressees, in order for it to be sucessfully implemented within the Member States' systems. In the shale gas Recommendation, the 'principles' are more evolved and substance is there, as the implementation within the Member States reflects. The Recommendation as such is

---

J. Black, 'Forms and paradoxes of principles-based regulation' *Capital Markets Law Journal* 3(4) (2008), 425–57.

[357] H. T. Demissie, 'Is beneficent regulation the new better regulation? Nano-regulation in the wake of the "new better regulation" movement' *Law, Innovation and Technology* 2(1) (2010), 115–49, at 144.

accordingly well drafted, but its practical impact depends on its implementation in the Member States.

Applying this reasoning to Collingridge,[358] the fact that the Union legislature refrained from additional legislative action at the beginning of the development of nanotechnology, did circumvent the dilemma at first and certainly promoted the practical application and development of the technology. Some scholars came to the conclusion that an 'incremental, reflexive and cooperative' approach is the way forward to regulate risks of nanotechnology in the near future.[359] However, should the existing regulatory framework on nanotechnology be deemed insufficient (which from today's perspective is not very likely to happen as both the knowledge base and the use of the technology have advanced), modifications of the existing regime would be difficult to establish in practice. Contrary to the case study on CCS, where regulation precedes the operation, law- and policy-makers in the EU decided to 'give the technology a go'; even if (regulatory) uncertainties existed. With regard to nanotechnology, it has been concluded that the underlying premises were that '[w]hile it is clear that more scientific information is needed, we need to act on what we know now, to enable industry to produce and market nanotechnology-enabled products that are as safe as possible',[360] as well as to stay competitive regarding the fast development of nanotechnology in the US. In addition, in the case of nanotechnology there was (and to a limited extent still is), 'increasing pressure from all sectors of society to determine how nanotechnologies are and will be regulated as they evolve'.[361] 'As such, a rare opportunity existed for stakeholders to actively participate in the development of a regulatory framework or frameworks for nanotechnologies.'[362] Shale gas is the second technology where this 'rare opportunity' arises, even if under changed regulatory competences at the European level, which add another layer of complexity to the regulatory design of the activity.

358 Section 2.1 above.

359 Marchant, Sylvester and Abbott, note 120 above, at 50; as well as Van Calster, Bowman and D'Silva, note 74 above, at 209.

360 Hansen et al., note 211 above, at 547.

361 Van Calster, Bowman and D'Silva, note 74 above, at 209.

362 Ibid.

## 5.2 Risk Management and the Underlying Principles

### 5.2.1 Precaution and prevention

When comparing the regulatory approaches taken for CCS and shale gas technology, it becomes apparent that the European Institutions, intervened at a much earlier stage in the process regarding CCS and that they used a stricter measure than for shale gas. In order to explain this, one might at first sight be inclined to take recourse to the precaution and prevention principles. It would seem logical to argue that the regulatory approaches taken towards the two technologies differ because they are based on different principles: for the CCS technology, the precautionary principle served as a basis for regulatory design,[363] whereas shale gas has been regulated under the heading of the prevention principle.

Consequently, the impacts associated with CCS are uncertainties rather than risks, as it is not only a question of likelihood, but about unknown unknowns to occur. Hence, in accordance with the precautionary principle, the Commission, rather than 'opt-out', introduced a regime of foresight which is living up to the best technique available. Shale gas on the other hand, as discussed above, is a modification of existing technologies which has been used for nearly a decade in the US, and as such, is not a new technology per se.[364] The prevention principle can explain the regulatory approach taken to date, even if this is not articulated as such. This would indeed be an easy way out, but this argumentation does not really fit in practice.

Indeed, CCS cannot be labelled a 'new' technology either. It equally is a modification of an existing technology, namely conventional oil and gas drilling. As such, 'enough technical experience on CCS is available around the globe (and also with European industry)',[365] although the 'existence of potentially serious risks [of CCS] with irreversible consequences has been mentioned. [However t]here is no consensus on the

---

363 As for example argued on the international level in M. Brus, 'Challenging complexities of CCS in public international law' in: Roggenkamp and Woerdman, note 13 above, 17–60, at 26ff.

364 See also Lucas, Watson and Kimmel, note 179 above, at 127f.

365 Ricardo AEA, et al., note 204, above, at 6.

existence of such risks.' [366] Indeed, the 'storage' part of the technology is still the centre of risk discussions,[367] especially regarding carbon leakage and related liability. This is more about regulatory uncertainty – meaning how to close this regulatory gap – rather than about scientific uncertainty though. More precisely, it is about *how*, meaning through which regulatory mechanism or tool, the technology can be best addressed, rather than *what* the impacts of the technology are as such. In short, while the use of the technology might be new, the underlying science is not. Hence CCS is also subject to prevention rather than precaution.[368] This goes hand in hand with the Eurobarometer findings on the acceptance of CCS which came to the conclusion that '[t]here was a lower level of caution about CCS technology among those who had greater knowledge about it'.[369] Whether this is also the case regarding nanotechnology is not clear.

The reason for the binding instrument on CCS is actually not one based on precaution or prevention reasoning. The differences in the regulatory approaches taken are rather attributed to the fact that the regulatory framework applicable to CCS has been adopted under the environmental competence prior to Lisbon and thus, also prior to the introduction of the energy title. It has been adopted at a point where the distribution between the competences in the environmental and energy area were (ironically) still much more straightforward and indeed clear. The lack of a binding measure for shale gas is attributed to the fact

---

366 European Commission, Proposal for a Directive on the geological storage of carbon dioxide and amending Council Directives 85/337/EEC, 96/61/EC, Directives 2000/60/EC, 2001/80/EC, 2004/35/EC, 2006/12/EC and Regulation (EC) No. 1013/2006, Brussels, 23.1.2008, COM(2008) 18, 2008/0015 (COD), C6-0040/08 at 4.

367 S. Bode and L. Dietrich, 'Regulating carbon capture and storage in the European Union: an economic and legal analysis' *Carbon Climate Law Review* 2 (2008), 196–201, at 200.

368 Interestingly, Mace et al. list the polluter pays and the precautionary principle (as well as the principle of intergenerational equity) as the key principles of international law applicable to CCS. This is especially striking as the prevention principle has a more established role in international law (see section 3.1 above). However, it seems the authors are confusing precaution and prevention, as in the following section they speak about 'risks' relating to CCS and precaution, whereas it should in that case have been rather 'uncertainties' and precaution or 'risks' and prevention. Mace, Hendriks and Coenraads note 184 above, at 254.

369 European Commission, 'Public awareness and acceptance of $CO_2$ capture and storage' (Special Eurobarometer, 2011), at 113.

that its regulation might not fall within the European competences at all.[370] Arguing *a contrario*, one can also conclude that a measure on CCS in the form we have today would not be so easy to adopt under the new set-up of energy and environmental competences. CCS is a cross-cutting energy and environmental issue which might also affect European energy policy.[371] One could further argue that (mandatory)[372] CCS might affect the Member States' right to determine their own energy resources, as CCS does not function in conjunction with all energy sources. It does however function with coal-fuelled power plants (and certain industry facilities). Admittedly, this argument is not very strong, but it cannot be disregarded completely.

One might wonder if the Lisbon Treaty changed the set-up of competences for nanotechnology, too. Legislation and policy on nanotechnology have been based on the environmental legal basis and subsequently on the Health and Safety Title (Article 153 TFEU). Since nanotechnology does not affect the energy area, at least not from a regulatory standpoint, there is no impact on the regulatory design post-Lisbon. The legal bases are essentially straightforward, much as they used to be.

Nevertheless the Commission (whether intentionally or not) might have learned lesson regarding Collingridge. Applying Collingridge to CCS might explain why the technology did not really fly until today: intervention might have come too early in the development and thus did not provide enough leeway for industry to further engage in the activity.[373] Normally the law has to keep up with technology. In the case of CCS this is arguably backwards, as the law is far ahead of CCS development.[374] As part of the Directive's review process, a stakeholder consultation has been carried out. It concluded that, according to the majority view of the stakeholders, there is not enough practical experience with CCS yet which would justify a complete revision of the Directive.[375] On the contrary; a revision of the Directive at this stage

---

370 See Chapter 2.

371 Which is not a problem as such.

372 As has been suggested in the past, see section 5.2.2 below.

373 Another important reason that CCS did not evolve as intended is the fact that the carbon price per t/$CO_2$ dropped from 40€ at the initial planning stage in 2008 to less than 5€ in 2011 due to the (European) financial crisis, and thus no investment incentives and consequently no business case existed.

374 McHarg and Poustie, note 190 above, at 249.

375 Ricardo AEA, et al., note 203, above, at 7f.

would lead to legal uncertainty[376] and increase the regulatory risk, eventually leading to 'additional delays'.[377] As there are no complete commercial CCS projects in the EU yet, the review should not take place before 2020, when completion of the first projects is expected.[378] The consultation further revealed that the Directive had 'no or [a] negative effect on progressing the speed of CCS deployment in Europe', and that the 'Directive has created an enabling framework for CCS rather than a mechanism to speed up or encourage CCS uptake'.[379] Compared to other jurisdictions, such as Canada for example,[380] CCS is said to be lagging behind:[381] none of the 12 CCS projects which were aimed at being realized in 2007 will be fully realized by the 2015 deadline.[382] The closure and post-closure obligations as established in the Directive, are perceived to be 'too *heavy a burden*' on the CCS industry, in addition to being 'not flexible enough given the often very different conditions'. Hence they are considered to be hampering CCS deployment'.[383] The Recommendation on shale gas includes vague and imprecise requirements on this point.

The case study of nanotechnology also provides some useful insights on risk management and the application of the precautionary principle for the shale gas regulatory debate.[384] Nanotechnology has been

---

376 For a discussion of the precise meaning of the term refer to Moses, note 24 above, at 250f.

377 Ibid.

378 Ricardo AEA, et al., note 204, above, at 3.

379 Ibid., at 8, as well as M. M. Roggenkamp and E. Woerdman, 'Looking beyond the legal uncertainties of CCS' in: Roggenkamp and Woerdman, note 13 above, 345–53, at 348.

380 See for more information, ICON, 'Carbon dioxide capture and storage – a Canadian clean energy opportunity' (2009), available at www.ico2n.com/wp-content/uploads/2010/07/ICO2N-Report_09_final.pdf; last accessed 05.07.2015.

381 Ricardo AEA, et al., note 203, above, at 9.

382 Ibid., at 12. This is of course not only due do the time and form of the drafting of the Directive, but external factors, such as the 2011 economic crises and the fact that there is no post-Kyoto mechanism to play an important role as well, see also, ibid.

383 Ibid., at 8f.

384 For a (rather superficial) comparison on the use of the precautionary principle regarding nanotechnology and a comparison between the US and EU interpretation of the principle, see Widmer and Meili note 329 above, at 255–9, as well as Kysar, note 72 above, at 3ff, and Marchant et al., note 105 above, at 252–6.

characterized as a clear case of precaution,[385] and the application of the precautionary principle as a principle for risk management and nanotechnology has been requested by NGOs,[386] Member States[387] and the Commission.[388] It also falls with the scope of the proportionality principle.[389] It is arguably an example of the 'correct' application of the principle.[390] A regulatory regime for nanotechnology has been established, proving that 'ostensible governability of emerging technologies [such as nanotechnology] should [and did] not prevent us from looking for ways to make such things more governable'.[391] Only a few voices[392] are of the opinion that none of the existing risk management methods work for nanotechnology.[393] Marchant et al. consider the precautionary principle as a risk management tool not workable on the grounds of its poor definition and limited practical value.[394] The bottom line reasoning for the application of the principle is the premise that in the past, in the presence of scientific uncertainties associated with a 'new' technology, one would 'sit back and call for more research'.[395] However, this attitude is no longer advisable today.[396] Nanotechnology is arguably one of the

---

385 C. Ganzleben, F. Pelsy and S. Foss, *Review of Environmental Legislation for the Regulatory Control of Nanomaterials* (Milieu Ltd. and AMEC Ltd., 2011), 17ff., as well as A. Grunwald, 'Nanoparticles: risk management and the precautionary principle' in: Jotterand, note 26 above, 85–102.

386 RPA and BiPRO, note 277 above, at 8.

387 Ibid.

388 European Commission, note 269 above, at 8.

389 Thus in essence submitting that the application of the precautionary principle should not result in a ban of the activity.

390 Miller and Scrinis are however of a different opinion and conclude that the NGOs involved in the regulatory design of the technology failed in promoting a precautionary approach. See Miller and Scrinis, note 243 above, at 430.

391 Stilgoe, note 281 above, at 2.

392 And in addition US voices, where the status and interpretation of the precautionary principle are quite different, see also section 3.1.

393 Marchant, Sylvester and Abbott, note 120 above, at 46f.

394 Ibid.

395 Hansen et al., note 211 above, at 547. As well as Grunwald, note 393 above, at 91.

396 Slightly differently in Marchant, Sylvester and Abbott, note 120 above, at 47. They argue that the precautionary principle, as well as any risk management principle, fails for the regulation of nanotechnology.

first new technologies to put the precautionary principle and its practical application to the test.[397]

Despite the (legal) scholarship, as well as some stakeholders promoting the precautionary principle as the risk managing principle, the regulatory framework on nanotechnology itself, is characterized by a 'general lack of express reliance on the precautionary principle' in most jurisdictions.[398] In the Union, this is only partially true. Whereas the Recommendation on the definition does not take recourse at all to the principle; the Code of Conduct promotes it as a key principle in governing nanotechnology.[399] The 2008 Communication on the regulatory aspects of nanotechnology stresses that: '[t]he scientific basis to fully understand all properties and risks of nanomaterials is not sufficiently available at this point in time';[400] and consequently that: '[w]here the full extent of a risk is unknown, but concerns are so high that risk management measures are considered necessary, as is currently the case for nanomaterials, measures must be based on the precautionary principle'.[401] Even if not referred to in all policy documents, the practical approach taken towards nanotechnology is a 'clear case of precaution' and of the 'right' application of the principle. The technology has not been banned because of its scientific uncertainties,[402] but it has been decided to regulate the technology 'as it evolves' and 'as needed'. As such, the precautionary principle from the beginning had a higher policy standing in the case of nanotechnology than in that of shale gas. In the shale gas regulatory debate, the Commission did not refer to the principle at all in the Recommendation, confusing the link between precaution and prevention in the accompanying documents.[403] This constitutes one of the fundamental differences in the regulatory approaches adopted for these technologies. Whereas the starting point was similar in the sense that some stakeholders called for the application of the precautionary principle, the application of the latter in the context of nanotechnology was never in doubt. The EU committed to a 'sustainable and responsible

---

397 As well as the first case where the traditional risk management principles (acceptable risk, cost–benefit analysis and feasibility) failed, according to ibid., at 44.

398 Hansen et al., note 211 above, at 541.

399 See also section 5.1.3.

400 European Commission, note 269 above, at 8.

401 Ibid.

402 Marchant, Sylvester and Abbott, note 120 above, at 51.

403 See also section 3.2 above.

approach'[404] to nanotechnology regulation, with the 'emphasis on precaution'.[405] Hence, nanotechnology was considered to be a 'real' new technology from its inception.[406] The case of shale gas is different as the application of the precautionary principle has not been straightforward. The scientific uncertainties and the 'newness' of the technology are key points in the regulatory debate.

Per the case of nanotechnology, setting aside the 'new' factor, an increased focus on risk management can remedy or at least soften the impacts of (possible) scientific uncertainties, at least as far as known unknowns are concerned.[407] The REACH Regulation, together with the application of a case-by-case approach through the pre-market approval system for nanotechnology substances,[408] including a risk assessment for each substance,[409] has proven to be the 'best possible framework for risk management'.[410] Nanomaterial substances are said to be 'different at the level of specifics (in that nano-silver has different properties from standard silver) but they are not different in kind'.[411] Hence, the risk management tools for 'normal' substances are equally applicable, which as a result render '[r]egulation that targets "nanotechnology", ... almost inevitably [to] become obsolete'.[412] For shale gas, such an approach is not directly transferable due to the very nature of the technology. Still, as outlined in Section 3.2, a shift from precautionary discussions to a discussion of proportionality concerns under the prevention principle would mean an assessment of similar tools. Indeed, the assessment on a case-by-case basis of shale gas activity seems to be the risk management tool which is undertaken right now in the individual Member States. However, a common risk management framework at a European level is needed to prevent a race to the bottom in terms of risk management and environmental protection standards.

---

404 Hunt, note 307 above, at 92f.

405 Ibid.

406 For example Woyke, note 26 above, at 14 and 16. A minority view in this regard is put forward by Stilgoe who argues that 'the novelty of nano remains slippery'. Stilgoe, note 281 above, at 2.

407 Van Calster, Bowman and D'Silva, note 74 above, at 226.

408 European Commission, 'Second Regulatory Review on Nanomaterials', Brussels, 3.10.2012, COM(2012) 572 at 5.

409 Ibid., at 11

410 Ibid.

411 Moses, note 18 above, at 16.

412 Ibid.

### 5.2.2 The regulatory principle of conservatism

The Commission has argued that it has employed the regulatory principle of 'conservatism' concerning the regulatory regime of CCS. The Commission itself explains this principle as follows:

> if the risk profile of a new activity (A) is comparable to that of an existing activity (B) already covered by a risk management framework, then that risk management framework is also adequate, effective and proportionate for managing the risks of activity A, and no consideration of further options is necessary. The risk profiles of two activities A and B are comparable if activity A does not present new risks that activity B does not, and it does not present significantly greater or lesser risks than does activity B.[413]

Regulation of the technology 'resulted hence in a conservative regulatory approach'.[414] So far, the principle of 'conservatism' is unique in European law-making. With the exception of the Impact Assessment for CCS, the principle of conservatism has not been included in any other policy documents of the Institutions, nor is it part of any better regulation or better law-making initiatives of the Commission.[415] It is also not included in the updated Better Regulation Package of May 2015,[416] which contains the Better Regulation Guidelines,[417] the Regulatory Fitness and Performance Programme (REFIT) plan and other policy documents.[418]

In response to the original problem of the role of science within law, risk regulation and management in general, the reasoning behind this principle is not helpful and complicates the discussion even further.[419] The explanation given by the Commission is far from clear, as it raises the questions of who is able to judge this and against which criteria? If science is to play a role in decision-making, decision-making would be trapped in a vicious circle, as proper risk management as interpreted from a principle of conservatism's perspective requires sound science. However, proper risk management is an option to deal with a degree of

---

[413] CCS Impact Assessment, note 250 above, at 23.

[414] European Commission, note 374 above, at 2.

[415] European Commission, 'European Governance: Better lawmaking', COM(2002) 275.

[416] European Commission 'Better regulation for better results – An EU agenda', Strasbourg, 19.5.2015, COM(2015) 215.

[417] European Commission, 'Better Regulation Guidelines. Commission Staff Working Document, Better Regulation Guidelines', Strasbourg, 19.5.2015, SWD (2015) 111.

[418] Ibid., at 110.

[419] See section 2 above.

uncertainty regarding a technology, thus in situations where sound science is absent. This leads to another question, namely how this principle interacts and relates to the precautionary and prevention principles?

The reference to the principle of conservatism as a regulatory approach for CCS had already been included in the first proposal of the Directive as presented by the Commission,[420] however it was absent in the preparatory documents by the working groups.[421] It has not been taken up, or commented on, by the other European Institutions (neither in the Council, nor in Parliament, or in the Economic and Social Committee) in the legislative procedure. The wording of 'conservative measures' was used for the first time in a study conducted on 'risk analysis of the geological sequestration of carbon dioxide' for the UK Department of Trade and Industry in 2003.[422] It has been taken up in a study by FIELD and ECOFYS,[423] which refers to 'conservative' measures mostly in relation to the precautionary principle,[424] however without explicitly defining the meaning of 'conservative'.

The Commission in turn seems to have taken up the concept of 'conservatism' in the CCS Impact Assessment and developed it further as the 'regulatory principle of conservatism'. It seems indeed that the principle has been 'invented' by the Commission and used in the context of CCS for the first and last time. At least consciously and with a direct

---

420 European Commission, note 374 above, at 5.

421 Such as for example in Second European Climate Change Programme, *Report of Working Group 3: Carbon Capture and Geological Storage (CCS)*, as agreed on 1 June 2006; as well as P. Zakkour, 'Choices for regulating $CO_2$ capture and storage in the EU – Task 2' (European Commission Discussion Paper, 2007); and ERM, 'Incentivising $CO_2$ capture and storage in the European Union – Task 3' (European Commission Discussion Paper, 2007).

422 M. Vendrig, J. Spouge, A. Bird, J. Daycock and O. Johnsen, 'Risk analysis of the geological sequestration of carbon dioxide' (DNV, DTI Report No. R246, 2003). The study is however not accessible any more. The author of this volume contacted both DNV Consulting and the authors to request a copy, without success.

423 C. Hendriks, M. J. Mace and R. Coenraads, *The Impacts of EU and International Law on the Implementation of Carbon Capture and Geological Storage in the European Union* (FIELD, ECOFYS, 2005), available at http://pdf.wri.org/ccs_impact_of_eu_law_on.pdf; last accessed 25.06.2015.

424 Ibid. They state that 'the precautionary principle requires that conservative measures be taken where scientific knowledge is not complete' at 40 and 55. General reference to conservative measures is further included at 14 and 50.

reference to the principle.[425] This is supported by the fact that reference to the principle in scholarship in the European regulatory context is largely missing.[426] Only one article has been identified regarding the use of this principle. It discusses uncertainty in the context of environmental impacts on the Hudson River, and thus examines the US regulatory context.[427] In the European context, one reference to the application of conservative measures in the context of CCS has been identified.[428] However, the article is written by the leading authors of the FIELD and ECOSYS study and hence, the conclusion that 'the precautionary principle requires that conservative measures be taken where scientific knowledge is not complete'[429] is not surprising.

Based on this principle, the Commission in its Impact Assessment reviewed the existing legislation on the individual elements of the CCS technology (capture, transportation and storage) separately. The Impact Assessment clarifies the underlying premises: '[a] conservative approach was taken, in the sense that the default option for regulating a CCS component was taken to be the existing legal framework that regulates activities of a similar risk (if one exists).'[430] The Commission came to the

---

[425] That it is arguably also applied without being mentioned as such regarding shale gas will be further established below.

[426] It has for example not been included in the overview of regulatory innovation in the European Union: Van Calster, note 41 above.

[427] J. R. Young and W. P. Dey, 'Uncertainty and conservatism in assessing environmental impact under §316(b): lessons from the Hudson River case' *The Scientific World* 2(1) (2002), 30–40. The article however employs a different, almost opposite definition of the principle. Whereas for the Commission the principle is used in the context of proportionate assessment and risk regulation regarding two activities associated with comparable risks, the authors put forward that the principle is used regarding 'highly uncertain' activities and more in the sense of the European interpretation of the precautionary principle. They for example put forward that: 'The precautionary approach explicitly recognizes the uncertainty of biological information and the imperfect ability of management policies to assure that biological targets are met. In recognition of this uncertainty, targets are set in a conservative manner …'. Further, they put forward that the more certainty is gained, the more the conservative approach of regulation is reduced; which in 'European speak' would mean a shift from precaution to prevention, see at 31, 34 and 35. Hence, the article was not considered to be helpful when analysing the European interpretation and meaning of the principle.

[428] Mace, Hendriks and Coenraads note 184 above, at 259.

[429] Ibid.

[430] CCS Impact Assessment, note 250 above, at 2 and 32.

conclusion that while for some elements e.g., regarding EIA and environmental liability, the existing legal framework can be used,[431] for other elements, especially relating to storage, a new legal framework was needed.[432] The former part (use of existing framework) is an articulation of the traditional risk management approach, meaning that the legislator usually tries 'to apply the current regulation to new technologies'.[433] This entails taking recourse to 'default rules' on ownership, liability and environmental legislation.[434] The regulatory principle of consesrvatism might as well be the application of the traditional risk management concepts in a new wrapping.

As such, the Commission decided not to amend the existing legislation relevant to the storage stage, such as the ETS Directive, the IPPC Directive (now IED) or existing waste legislation, as these were not appropriate to also deal with storage.[435] Rather, and indeed to a certain extent, contradictorily, as the CCS Directive in the end amends certain existing legislation, the Commission came to the conclusion that 'one single framework will be sufficient',[436] and that the CCS Directive managed to create a climate change 'mitigation option, and that it is done safely and responsibly.[437]

What is striking is that the Impact Assessment on CCS does not include any reference to the prevention principle only a few references to

---

431 European Commission, note 374 above, at 2.

432 CCS Impact Assessment, note 250 above, at 2.

433 M. Haenchen, P. Cocco and U. Kastrup, *Use of the Underground – Background Document for EEA Scientific Committee Seminar* (EEA, Doc. EEA/SC/61/sem-02rev.1, 2015), at 20. Even if the current legal framework has not been drafted with CCS in mind. See also Purdy and Havercroft, note 11 above, at 353, as well as H. de Coninck et al., 'Is $CO_2$ capture and storage ready to roll?' *Journal for European Environmental & Planning Law* 4(5) (2007) at 407f, also for an overview of existing Directives and concerns regarding their applicability.

434 Haenchen, Cocco and Kastrup, ibid., at 20.

435 European Commission, note 374 above, at 5.

436 Ibid., at 7.

437 Ibid., at 2. This is upheld by Roggenkamp et al. who conclude that: 'the CCS Directive provides ample opportunities for CCS. We have found no outspoken inconsistencies or blunt mistakes in the CCS Directive. In principle, it provides a solid legal basis …'. Roggenkamp and Woerdman, note 388 above, at 350.

precaution in other contexts than risk regulation.[438] The Directive itself does not include any reference to these principles. The principle of conservatism is however discussed and used as reasoning for regulatory design decisions in multiple instances.[439] One would be right to wonder why the traditional risk governance and environmental principles of precaution and prevention were not deemed appropriate and/or applicable in establishing a governance approach for CCS.

The lack of references can be explained by the fact that CCS, as established above, is also not an actual 'new technology'. Neither of these principles are at the forefront of reasoning and drafting of regulatory intervention. Indeed for shale gas (also not considered to be a new technology) the prevention and precautionary principles do not seem to be of significant importance for the Commission,[440] which supports this interpretation. Applying the parameters of the conservatism principle to shale gas, 'the risk profile of a new activity A [shale gas] is comparable to that of an existing activity B [conventional oil and gas extraction] already covered by a risk management framework'. The risk management framework is then automatically also:

> adequate, effective and proportionate for managing the risks of activity A [shale gas], and no consideration of further options is necessary. The risk profiles of two activities A [shale gas] and B [conventional oil and gas extraction] are comparable if activity A [shale gas] does not present new risks that activity B [conventional oil and gas extraction] does not, and it does not present significantly greater or lesser risks than does activity B [conventional oil and gas extraction].

The regulatory principle of 'conservatism' has thus arguably also been applied in the shale gas regulatory debate, however without being mentioned explicitly. With regard to shale gas, the existing legislative framework has been assessed both at Member State and European levels to see whether shale gas can be addressed by the existing framework.[441] It is found that '[b]oth general and environmental legislation of the Union apply to hydrocarbon exploration and production operations involving

---

[438] Once in relation to costs and failure to invest, CCS Impact Assessment, at 47, and once regarding the risk of leakage, CCS Impact Assessment, note 250 above, at 67.

[439] See ibid., at 2, 25, 32, 41 and 54.

[440] See section 3.2 above on the use of the principle in the Commission documents on shale gas.

[441] Such as Ballesteros, Pelsy and Reins, note 63 above.

high-volume hydraulic fracturing'.[442] Only some environmental aspects, more precisely strategic planning, underground risk assessment, well integrity, baseline and operational monitoring, capturing methane emissions and disclosure of information on chemicals are not addressed in the existing framework, due to the fact that they preceded the development of the technology.[443] The consequence however, was not to adopt a specific Directive on shale gas, which would have consisted of a 'conservative' approach; but the adoption of minimum principles, ergo a more 'liberal' approach. This is striking, especially considering the substantial public debate associated with shale gas and the relatively unaware public regarding CCS, as further explained below.

Deriving from the definition of the conservatism principle as established by the Commission along with the way it is used to justify the legislative framework for CCS, the conservatism principle links with the 'traditional' precautionary and prevention principles at the proportionality stage.[444] The Commission itself refers to the 'interests of proportionate assessment', and establishes the criteria of adequateness, effectiveness and proportionality as the reference point of a risk management framework.[445] This means that the principle of conservatism can in fact be considered a 'correction to the precaution and prevention principles', in the sense that the traditional principles are associated with either 'new' technologies and scientific uncertainties in the former case (precaution), and/or known risks and likelihoods in the latter case (prevention). Table 3.4 illustrates these approaches.

---

[442] Recommendation on Shale Gas, Preamble 7.

[443] Ibid., Preamble 8.

[444] Both principles are subject to proportionality considerations, as established in section 3.1 above. In addition, Lierman puts forward that 'The principle of proportionality is a principle that helps to balance conflicting principles and interests.' S. Lierman, 'Law as a complex adaptive system: the importance of convergence in a multi-layered legal order' 21 *Maastricht Journal of European and Comparative Law* 4 (2014), 611–629, at 629.

[445] Impact Assessment at 23.

*Table 3.4 Relationship between the principles of precaution, prevention and conservatism*

| Circumstance | Regulatory principle |
|---|---|
| Uncertainty: unknowns and unquantifiable | Precaution |
| Risk: known unknowns and likelihood | Prevention |
| Risk: known and quantifiable | Conservatism |

The conservatism principle arguably comes into play for technologies where there is already a certain legislative framework available; the technology is only a modification of an already existing technology and hence not all aspects need to go through regulatory scrutiny on the grounds of proportionality considerations. This is further supported by a finding of the FIELD and ECOSYS study: with reference to the 2003 DTI study, they conclude that 'the generic risk assessment included a relatively high degree of conservatism and that there is significant scope for the individual risks associated with actual applications to fall within acceptable limits'.[446] The regulatory principle of conservatism is arguably the traditional risk management approach in a new package, and as such a regulatory practice or a 'risk assessment tool'[447] rather than a defined concept or principle.[448]

### 5.2.3 Best available techniques

Best available techniques (BAT)[449] Reference Documents are one tool for risk management.[450] For CCS, no horizontal BREF (Reference

[446] FIELD, ECOFYS, note 431 above, at 55.

[447] The 'risk management tool versus principle discussion' is not a new one and has been debated regarding the ALARA approach in nuclear power for example. See also Lierman and Veuchelen, note 24 above, at 99.

[448] As has been also argued for the 'command and control' approach, Van Calster, note 41 above, at 294.

[449] The concept of BAT was first introduced by the IPPC Directive. According to Article 2(12) of the Directive:

> 'best available techniques' means the most effective and advanced stage in the development of activities, and their methods of operation which indicate the practical suitability of particular techniques for providing in principle the basis for emission limit values designed to prevent and, where that is not practicable, generally to reduce emissions and the impact on the environment as a whole:
>
> (a) 'techniques' shall include both the technology used and the way in which the installation is designed, built, maintained, operated and decommissioned;

document containing information on the best available techniques) exists,[451] however the activity is included in the updated BREF on Large Combustion Plants.[452] Furthermore, the CCS activity is included in the update on the BREFs on Cement and Lime Manufacturing and Mineral Oil and Gas Refineries.[453] Shale gas is generally regulated through the Mining Waste Directive (MWD),[454] meaning that it is subject to management in specialized facilities in compliance with specific rules and the BAT (Article 4 MWD). However, as discussed above, the MWD provides a general framework for mining waste, but does neither address the special characteristics of the shale gas activities, nor the characteristics of the waste associated therewith. Tellingly, until now no *specific* BREF has been in place for the shale gas activity. The existing BREF on the Management of Tailings and Waste-Rock in Mining Activities,[455] adopted in 2009, is currently under review under the lead of the SUSPROC (Sustainable Production and Consumption) project by the Joint Research

---

(b) 'available techniques' means those developed on a scale which allows implementation in the relevant industrial sector, under economically and technically viable conditions, taking into consideration the costs and advantages, whether or not the techniques are used or produced inside the Member State in question, as long as they are reasonably accessible to the operator;

(c) 'best' means most effective in achieving a high general level of protection of the environment as a whole.

[450] On the role of BAT and BREFs in the European (risk) regulatory process, refer to Lee, note 75 above, at 119ff.

[451] Despite it having been asked for by several institutions, such as the European Parliament Committee on the Environment, Public Health and Food Safety in its draft opinion on Supporting Early Demonstration of Sustainable Power Generation from Fossil Fuels (2008/2140(INI)), 23.7.2008.

[452] Reference Document on Best Available Techniques for Large Combustion Plants, July 2006. This is currently under review: Best Available Techniques (BAT) Reference Document for the Large Combustion Plants, Industrial Emissions Directive 2010/75/EU, Draft 1 (June 2013). See also De Graaf and Jans, note 13 above, at 180.

[453] See for example Zakkour, note 429 above, at 22.

[454] Directive 2006/21/EC on the management of waste from extractive industries and amending Directive 2004/35/EC, [2006] OJ L 102/15.

[455] European Commission, BREF on Management of Tailings and Waste-Rock in Mining Activities (2009), available at http://eippcb.jrc.ec.europa.eu/reference/BREF/mmr_adopted_0109.pdf; last accessed 25.06.2015.

Centre's Institute for Prospective Technological Studies (JRC-IPTS) of the European Commission. Results are not expected until 2017.[456]

In terms of risk management, the issuance of BAT bypasses the step of risk analysis and jumps directly to the phase of risk reduction.[457] This is certainly appealing when confronted with 'new' technologies associated with a vast amount of scientific uncertainties, such as nanotechnology and shale gas. The problem with applying BAT to technologies which are surrounded by uncertainties, is that these technologies might well be over- or indeed under-regulated, as it is not possible to determine the exact impacts beforehand.[458] Applying this reasoning to shale gas, one can conclude that whereas the drafting of specific BAT is a correct application of the prevention principle and an important, correct step in terms of risk management, one should refrain from using this approach as an opt-out mechanism from the risk analysis phase.

### 5.2.4 Interim conclusion

Comparing the regulatory approaches taken towards nanotechnology, CCS and shale gas, one can conclude that the creation of a new (tailor-made) regulatory regime is not necessary, not even when it comes to the regulation of an actual new technology (such as nanotechnology). Furthermore, a regulatory framework drafted prior to a technological innovation can still be enough to comprehensively regulate a technology; likewise, under precautionary considerations. Moreover, even if shale gas were considered to be an actual new technology, or indeed to fall under the regulatory scope of the precautionary principle rather than the prevention, this would not necessarily and automatically result in a stricter, more comprehensive or more coherent regulatory regime. As the case of nanotechnology shows, 'recourse to the precautionary principle does not necessarily mean adopting final instruments designed to produce legal effects. A wide range of activities or measures can be used,

---

456 For more information visit: SUSPROC, 'Waste and Recycling' available at http://susproc.jrc.ec.europa.eu/activities/waste/index.html; last accessed 25.06.2015.

457 Marchant, Sylvester and Abbott, note 120 above, at 45. Also a general problem with the practical application of the precautionary principle. See Hansen and Tickner, note 134 above, at 18.

458 See also Moses, note 3 above, at 601f.

like legally binding measures, initiation of research projects or recommendations.'[459] This statement has to be read against the backdrop that the Commission at the beginning of nanotechnology development had the mission to push the technology. As such, there are two types of application of the precautionary principle towards new technologies. First, the Commission wants to do something and uses the precautionary principle as an excuse to act. In this scenario, the precautionary principle only comes into play at a later stage in the decision-making process, as a justification for an action. This application was also the case for nanotechnology. A second, and arguably 'better' application of the principle, the precautionary principle forces someone, i.e., mostly the Commission, to act (see also Table 3.4 above). The principle is the starting point for a decision-making process and debate. This application is yet to be the case in EU regulatory history. Concerning shale gas, the application of the precautionary principle will not be the prevailing one, as the uncertainties involved in the technology are not as prevalent.[460]

## 5.3 The Business Case

This section compares the business case for shale gas and CCS, as both technologies are very capital-intensive.[461] The need for capital for nanotechnology is not an issue[462] as nanotechnology has been referred to as 'a major driver of economic growth in the first part of the new millennium'.[463] Its business case is straightforward. Therefore, the technology has been excluded from the comparison.

Regarding the regulation of CCS, law- and policy-makers in the Union had a dual objective from the very beginning: on the one hand the management of environmental risks, and on the other hand addressing commercial barriers to development, thus the need for creating incentives

459 European Commission, note 269 above, at 8.

460 See section 3.2 for a detailed discussion.

461 For the predictability of the economics and costs involved in CCS, refer to Coninck et al., note 441 above, at 411f.

462 In addition, the overall support for the development of nanotechnology (especially regarding financing, promotion of the technology, but also political support) from both governments, but also the industry sector is strong. See Miller and Scrinis, note 243 above, at 409, as well as Woyke, note 26 above, at 14.

463 Kysar, note 72 above, at 243.

for development.[464] Investors not only look for legal certainty and stable legal framework for their investments, but also for economic incentives.[465] Regarding CCS, policy-makers departed from the presumption that it is not sufficient to leave it up to the market to invest in the technology,[466] and that incentivization of CCS is crucial for its development.[467] With the inclusion into the EU ETS, the Commission believed it had established such an incentive,[468] alongside Commission funding programmes[469] and Member States' (funding) initiatives.[470]

The Commission clearly stated that '[t]he EU is committed to supporting CCS both financially and with regulatory steps'.[471] In the shale gas regulatory debate, the first form of support is not an issue and has not been mentioned in any policy document. The regulatory debate is solely focused on environmental protection. The Recommendation as it stands, certainly does not create an investor-friendly framework, however it might create some legal certainty (as much as a non-binding instrument can, that is), which arguably influences investors.

In CCS, the regulatory framework, legislative history and debate started from the general positive presumption of making the technology work: the focus is not so much 'if', but 'how' and under which enabling

---

464 European Commission, note 440 above, at 2 and 4. The Commission Communication even included three objectives of a EU regulatory framework: 'ensure the environmentally sound, safe and reliable operation of CCS activities; remove unwarranted barriers to CCS activities in current legislation; provide appropriate incentives proportionate to the $CO_2$ reduction benefit': European Commission, 'Communication on Sustainable power generation from fossil fuels: aiming for near-zero emissions from coal after 2020', Brussels, 10.1.2007, COM(2006) 843 at 8.

465 Roggenkamp and Woerdman, note 387 above, at 349.

466 Impact Assessment, at 19, as well as Grevers and Luten, note 187 above, at 11.

467 Purdy and Havercroft, note 11 above, at 364.

468 For a detailed discussion, refer to E. Woerdman and O. Couwenberg, 'CCS in the European Emissions Trading Scheme' in: Roggenkamp and Woerdman, note 13 above, 97–125.

469 See for example the NER300 programme, further discussed in section 5.1.3 above.

470 For a case study on CCS development framework in the UK, please refer to E. Peters, 'What can be done to help the development of a UK CCS industry?' *Journal for European Environmental & Planning Law* 4(5) (2007), 393–401.

471 European Commission, Communication on the Future of Carbon Capture and Storage in Europe, Brussels, 27.3.2013, COM(2013) 180 at 3.

framework and conditions.[472] This can already be derived from the four possible options to act regarding CCS; the conclusion that 'Option 0 CCS not enabled' would have negative impacts on employment and competitiveness.[473] The other options consisted of e.g., making CCS mandatory after a certain time and providing investment subsidies.[474] Additionally, the impact assessment already includes suggestions for $CO_2$ storage site selection (Annex III) and a methodology for the carrying out of an EIA (Annex V). Whether shale gas needs an EIA at all is still subject to debate. Hence, the discussion surrounding shale gas is different. Due to the introduction of the energy title which clearly states that the choice of energy supply is at the discretion of the Member States, the focus cannot be on 'if at all'. The European level can at best, if a Member State decides to develop the technology, lay out conditions focused on environmental aspects, and no more. As shown in Chapter 2, even the competence on environmental issues is not straightforward.

Financial incentives associated with the shale gas case are quite clear. Commercial profitability is not an issue; it is clear that shale is profitable.[475] That is not necessarily the case for CCS: without financial support, as well as regulatory clarity, CCS cannot fly.[476] In the end, due to low carbon prices among other complications,[477] CCS did not prove a

---

472 For an overview and discussion of economic considerations associated with CCS, see Woerdman and Couwenberg, note 477 above, at 99ff.

473 CCS Impact Assessment, note 250 above, at 85.

474 Ibid. This was however rejected as it would have been feasible only at substantial additional cost. See also De Graaf and Jans, note 13 above, at 180.

475 See also Triple E Consulting, *Economic Impacts of Shale Gas in the Netherlands* (2014). The report includes calculations for the Netherlands, Europe and the US. It comes to the conclusion that there is a business case for shale gas certainly in the US. However due to a lack of data for the European and Dutch examples, only estimations can be made. It concludes that a business case for shale gas is possible, however it depends on the amount of resources and social costs.

476 This is also stressed by Mace et al. who state that:

> In a context of rising greenhouse gas emissions, where CCS is very costly and geological storage shown to be safe, a variety of economic incentives may be needed to encourage the uptake of CCS technology and to increase its cost-efficiency through large-scale application.

Mace, Hendriks and Coenraads note 184 above, at 260. A discussion of several forms of incentives, especially regarding tax law can be found in J. N. Bouwman and I. J. J. Burgers, 'The law as a barrier or tool for promoting CCS' in: Roggenkamp and Woerdman, note 13 above, 321–44.

477 See also European Commission, note 480 above, at 16.

viable technology as it failed to provide sufficient incentives for its investors.[478] The market and regulatory failure of CCS in the Union illustrates that supporting schemes for new technologies is a limited response and only sensible when the support is small and economically inconspicuous.[479] This has not been the case for CCS, where large-scale sums from governments for the financing of demonstration projects have been predominant in trying to make the technology work in practice,[480] combined with a detailed regulatory framework. As has been concluded however, a 'simpler and more committed regulatory framework would [arguably] prove both cheaper and more effective in the long run'.[481]

## 5.4 Addressing the Public Debate

There are generally three overall motivations for public participation. First, a normative motivation, in the sense that participation is 'a good thing in itself'; second, an instrumental motivation, namely that participation is a tool to build trust and can smooth the implementation on decisions; and lastly, substantive participation with the motivation to produce better decisions.[482] The instrumental motivation behind public participation is the motivation which drove participation in the past,[483] and is also the prevailing one regarding CCS, shale gas and nanotechnology.

As previously discussed, the initial problem regarding CCS[484] was a problem of public awareness[485] rather than public acceptance, as is the

[478] See also Langlet, note 14 above, at 198; as well as A. Boute, 'CCS under the project-based Kyoto mechanism' in: Roggenkamp and Woerdman, note 13 above, 61–95 at 74. For a discussion of different instruments providing efficient incentives, such as a carbon tax, double credits, a closure of the cost gap and mandatory CCS, please see Woerdman and Couwenberg, note 477 above, at 106–23.

[479] C. Mitchell and B. Woodman, 'Regulation and sustainable energy systems' in: R. Baldwin, M. Cave and M. Lodge (eds) *The Oxford Handbook of Regulation* (Oxford: Oxford University Press, 2010), 572–89.

[480] Havercroft and Macrory, note 11 above, at 301.

[481] McHarg and Poustie, note 190 above, at 274.

[482] D. J. Fiorino, 'Citizen participation and environmental risk: a survey of institutional mechanisms' *Science, Technology & Human Values* 15(2) (1990), 227–8, as also used by Stilgoe, note 281 above, at 5.

[483] Stilgoe, ibid., at 5.

[484] This is not the case any more. In Europe, two demonstration projects have been delayed or abandoned due to public opposition. See: I. Havercroft, R. Macrory and R. B. Stewart, 'Introduction' in: Havercroft, Macrory and

case with shale gas. According to a Special Eurobarometer study from 2011 on public awareness and acceptance of CCS, only 10 per cent of more than 13,000 participants said 'they had heard of CCS and knew what it was', while 18 per cent 'had heard of it but did not really know what it was'.[486] Awareness is especially low in Spain, Poland, the Czech Republic and France where more than 75 per cent of the respondents claimed that they had not heard of CCS.[487] One possible explanation for these statistics is that CCS is not necessarily visible (nor is nanotechnology for that matter), and only adds a few components to existing facilities. One might argue that this is the same for shale gas as the technology operates in large part underground. However, shale gas is not as inconspicuous to the public eye as CCS. A variety of factors including increased truck movement(s), impacts on water resources, venting and flaring installations, visible impacts from boreholes and kilometres of well equipment are visible to the public eye. However, with nanotechnology, even if it is invisible, people (consumers) still have a choice on whether to buy and/or use a product containing nanomaterial (assuming the product has been labelled correctly.[488] The situation with shale gas and CCS is different. Once the facilities are constructed, the local residents cannot 'avoid' the technologies.

In addition to the lack of awareness among the population regarding the aforementioned technologies, or indeed rather linked to it, is the public's perception of risks relating to CCS.[489] Most questions related to the fact that 'the technology was insufficiently mature to be mandated, and that the implications for the energy mix were unpredictable', [490] thus not to environmental and human health issues directly, as is the overarching point of concern regarding shale gas. That risk perception is lower can also be derived from the fact that 60 per cent of respondents agreed that 'capturing and storing $CO_2$ should be compulsory when building a

---

Stewart, note 11 above, 1–4, at 4, as well as Ashworth and Cormick, note 263 above, at 251.

485 Fouillac, note 186 above, at 139.

486 European Commission, note 280 at 10.

487 Ibid., at 75.

488 For a detailed discussion of the labelling and certification regarding nanotechnology, please refer to Weidl, Klein and Zöllner, note 222 above.

489 See also note 44412 above, at 403. Not even NGOs, while being sceptical, are completely against the technology, ibid., at 407.

490 European Commission, note 374 above, at 2; as well as Impact Assessment, at 10.

new coal-fuelled power plant'.[491] An additional indication is the fact that the classical 'NIMBY' (Not In My Back Yard) syndrome, which presents a sizeable hurdle for shale gas is not of great relevance to CCS, with the exception of a few individual cases.[492] Generally, the awareness is rather low even in countries with major EU co-financed CCS projects.[493] The Eurobarometer study revealed that 61 per cent of the people would be 'worried', and 24 per cent even 'very worried', if an underground storage site for $CO_2$ were to be located within 5 km of their home.[494] Indeed, the general acceptance of CCS can differ greatly from acceptance of a specific project in one's immediate neighbourhood.[495] Offshore storage projects in the UK,[496] the Netherlands and Italy have gained the acceptance of the public.[497] In Spain, initial public opposition to a site was overcome by a 'dedicated information and engagement campaign'.[498] An exception to the general perception of low risk regarding CCS is the public debate which took place in Germany. There, public opposition was mentioned as one of the key reasons for the delay in the transposition of the CCS Directive, eventually leading to a halt of the German pilot projects.[499]

Nevertheless, the Commission stressed correctly 'that public acceptance should not be assumed without prior assessment'.[500] And indeed, the review of the CCS Directive indicated that the design and content of

---

[491] European Commission, note 280 at 13.

[492] This is essentially due to the fact that the 'visible part' of CCS facilities are mostly connected to already existing industry facilities, and thus do not impose a lot of additional 'visible' impacts. An additional reason is that also the awareness of (possible) implications is not that high. For more on NIMBY and CCS: S. M. Forbes et al., 'Regulatory requirements for public engagement around carbon dioxide capture and storage demonstrations' in: Havercroft, Macrory and Stewart, note 11 above, 265–73, at 275.

[493] European Commission, note 280 at 77.

[494] Ibid. 92.

[495] Haenchen, Cocco and Kastrup, note 441 at 405.

[496] For a case study on public participation in CCS planning in the UK, please refer to L. Meyric and N. Westaway, 'Public participation in UK CCS planning and consent' in: Havercroft, Macrory and Stewart, note 11 above, 277–96. The authors conclude that the adequacy and effectiveness of the public involvement in the UK largely remains uncertain. For an analysis of the general approach taken towards the regulation of CCS in the UK, refer to: McHarg and Poustie, note 190 above.

[497] European Commission, note 480 above, at 18.

[498] Ibid.

[499] Ibid.

[500] Ibid.

the Directive as such is a problem for the few existing public opposition cases. More concretely, industry stakeholders believe that the CCS Directive 'sends too many signals about "uncertainties and lack of safety" of CCS, and thus that the Directive is indirectly hampering public acceptance';[501] as well as that public opposition and concern are best addressed at the local and individual project levels. This is also a concern regarding the shale gas debate. The Recommendation in its Preamble clarifies that:

> *a set of rules* would *level the playing field* for operators, and improve investors' confidence and the functioning of the single energy market. Clear and transparent rules would also help *alleviate public concerns*, and *possibly opposition* to shale gas development. (Preamble 9, emphasis added)

Again, the question is whether a non-binding Recommendation can really be called a 'set of rules', and furthermore, whether a 'level playing field' can be established in this way. As has already been concluded with respect to CCS as well: 'the more binding the CCS policy proposed, the more the opinions are likely to differ'.[502] However, the CCS experience shows that binding regulation in the form of a Directive, that is, some minimum form of legal certainty and harmonization,[503] does not necessarily have a positive influence on public acceptance of a project. On the contrary, according to the draft Recommendations for the review of the CCS Directive:

> both industry and one NGO … feel that the over-prescriptive requirements of the Directive and Guidance Documents are a factor in the slow progress of CCS, both directly and indirectly by *fuelling public concerns*.[504] (Emphasis added.[505])

The stakeholders call for more 'flexibility at the MS [Member State] level [which] is key to further development of CCS',[506] something that the Recommendation, already because of its form, no doubt grants.[507]

---

[501] Ricardo AEA, et al., note 203, above, at 8.
[502] Ibid., at 7f.
[503] See also De Graaf and Jans, note 13 above, at 157.
[504] Ricardo AEA, et al., note 204, above, at 9.
[505] See also Haszeldine, note 194 above, at 18.
[506] Ricardo AEA, et al., note 203, above, at 8.
[507] This is for example stressed several times in the Impact Assessment on Shale Gas. See European Commission, note 161 above, at 19, 42 and 44.

Comparing shale gas and nanotechnology, lessons can also be learned from the public debate surrounding both technologies.[508] The underlying problem here is the consideration that a premature public opinion on a certain technology regarding the perceived risks and benefits can hinder the development of the technology.[509] Whereas in the case of nanotechnology the European Institutions created an overarching strategy on 'Communicating Nanotechnology'[510] and used awareness raising campaigns,[511] the communication and public debate surrounding shale gas technology has been characterized by public fears and alarmist newspaper headlines from the beginning. Admittedly, the nanotechnology stakeholders only managed to keep the discussion neutral for a brief period of time.[512] The resulting negativity has been 'nano-phobia-phobia': the regulators' fear of public backlash and opposition to the technology.[513]

Initially, certainly also due to the cooperative industry/NGO approach, the technology, unlike shale gas, was not subject to excessive media scrutiny and associated safety concerns.[514] These emerged slowly and in low numbers.[515] In 2006, a cleaning product called 'Magic Nano' caused 100 illnesses and made some headlines. The incident resulted in a media storm, and generated a call for a worldwide ban of nanotechnology. The storm ended however, when it was revealed that the product did not

---

[508] For a detailed review of the civil society actors within the nanobiotechnology (regulatory) debate, please refer to Plows and Reinsborough, note 243 above.

[509] Marchant, Sylvester and Abbott, note 120 above, at 47. See also Birnbacher, note 26 above, at 177. See also Ashworth and Cormick, note 263 above, at 252.

[510] European Commission 'Why, to whom, saying what and how?' (2010), available at ftp://ftp.cordis.europa.eu/pub/nanotechnology/docs/communicating nanotechnology_en.pdf; last accessed 25.06.2015.

[511] For example the introduction of the 'nanochannels' webpage, as part of an FP7 project on NANOCHANNELS 'Engaging European stakeholders in debating Nanotechnology issues using a range of media channels', available at www.nanochannelsfp7.eu; last accessed 25.06.2015.

[512] For the main concerns raised by NGOs, see Miller and Scrinis, note 243 above, at 411.

[513] A. Rip, 'Folk theories of nanotechnologists' *Science as Culture* 15(4) (2006), 349–65, at 350 and 357ff.

[514] See also Miller and Scrinis, note 243 above, at 409, as well as Grunwald, note 393 above, 87f.

[515] B. M. Hutter, *Anticipating Risks and Organising Risk Regulation* (Cambridge: Cambridge University Press, 2010) at 6, as well as Miller and Scrinis, note 243 above, at 411.

actually include nanomaterials.[516] In general, the neutral discussion was short-lived, although it was ultimately the unbalanced media attention paid to nanotechnology that raised fears and influenced people's risk perception regarding the technology.[517]

In 2012, eight years after the first European efforts on nanotechnology, it was said that the 'public knowledge and understanding regarding nanotechnology is currently lagging far behind the rapid pace of nanotechnology development',[518] and that a 'majority of people [have] never ... heard of nanotechnology'.[519] NGOs' attempts to draw public attention to the issue of nanotechnologies and related risks began slowly and arguably with little success,[520] but were manifold in the end.[521] As a consequence, the Commission launched a campaign to raise 'ethical, legal and social aspects of nanotechnology applications'.[522] The nanotechnology stakeholders, at least regarding communication, public participation and debate, as well as consumer concerns, seemed to have learned from the experience with the introduction of biotechnology in the Union.[523] The introduction of this technique was, and largely still is, characterized by public opposition and rejection of the technology.[524] The case of nanotechnology was also an 'experiment in democracy',[525] as a bottom-up approach of public engagement was witnessed for the first

---

516 Marchant, Sylvester and Abbott, note 120 above, at 48; as well as K. W. Abbott, D. J. Sylvester and G. E. Marchant, 'Transnational regulation of nanotechnology: reality or romanticism?' in: Hodge, Bowman and Maynard, note 41 above, 525–44, at 527.

517 Marchant, Sylvester and Abbott, note 120 above, at 48; as well as Grunwald, note 393 above, at 88 and Ashworth and Cormick, note 263 above, at 256f.

518 European Commission, note 520 above.

519 Ibid.

520 Miller and Scrinis, note 243 above, at 409, as well as Van Calster, note 34 above, at 70.

521 See for an overview of NGO coordination, regulatory and governance proposals and activities (not only but also on the European level): Miller and Scrinis, note 243 above, at 412–19.

522 Ibid.

523 Mandel, note 98 above, at 7, as well as Van Calster and Bowman note 91 above, at 278 and Ashworth and Cormick, note 263 above, at 255.

524 Several academic articles have been written on the issue, see for example: G. Gaskell, 'Agricultural biotechnology and public attitudes in the European Union' 3 *AgBioForum* 3(2/3) (2000), 87–96, as well as M. Bucchi and F. Neresini, 'Why are people hostile to biotechnologies?' 304 *Science* 304(5678) (2004), 1749.

525 Stilgoe, note 281 above, at 4.

time.[526] Nanotechnology in the UK in particular, tried to gain public confidence by introducing participatory mechanisms[527] and 'vast road shows'.[528] As Van Calster concludes, this 'evidently cannot be repeated for every new technology that emerges'.[529] Early participation of the public is thus identified as a key requirement for risk management.[530] In the case of shale gas, it is doubtful whether such a 'road show' would make a difference at all. The situation is arguably worse than it was regarding nanotechnology and biotechnology; because the public debate is essentially characterized by fears of environmental harm, negative impacts on human health and the classic 'NIMBY' reaction. Protests are ongoing in various Member States, such as Bulgaria, Germany and the UK Risk communication and integration of the public into decision-making at European level failed and 'reproducible' science, as established further above,[531] is rare. In addition, the general public is often focused on the type of potential harm and not on the probability of its occurrence.[532]

This holds especially true for shale gas. As established in Section 3.2, a shift from precautionary to preventive tools is necessary in order to improve risk management and communication on shale gas technology. The Commission seems to have forgotten the lessons learned in risk communication regarding nanotechnology and has to start anew. This is especially disturbing as the public is normally guided by the governance of science rather than the (quality of) science as such.[533] In addition, 'NIMBY' is certainly a predominant factor, regarding shale gas. Moreover, a premature opinion is established; the fact that this technology is only regulated through a non-binding Recommendation certainly does not increase public confidence and trust in either the technology or the European legislature as such.

---

526 Ibid.

527 See also Plows and Reinsborough, note 243 above 142.

528 Van Calster, note 34 above, at 70.

529 Ibid.

530 Renn and Grobe, note 163 above, at 502ff.

531 See further section 3.2 above.

532 Marchant, Sylvester and Abbott, note 120 above, at 48.

533 J. Wilsdon, B. Wynne and J. Stilgoe, *We Need to Infuse the Culture and Practice of Science with a New Set of Social Possibilities – The Public Value of Science – Or How to Ensure that Science Really Matters* (London: Demos, 2005) at 29.

Moreover, the regulatory process leading up to the two Recommendations on nanotechnology was slow[534] and arduous.[535] As part of this process, a public consultation on transparency for nanomaterials on the market has been carried out. The consultation, similar to the CCS consultation, revealed the concern of the industry sector that, '[a]ccording nanomaterials different treatment (e.g., providing consumers with information on products containing nanomaterials even if safe, as use has been demonstrated) will result in a stigmatization of nanomaterials, with a negative effect on consumer trust'.[536] Therefore, there should not be a specific legal framework for the technology. Rather it should be regulated 'in the same way as other substances',[537] as it is feared that further requirements would 'negatively impact the competitiveness and innovative capacity of the chemical industry'.[538] The same argumentation is applicable not only regarding CCS, but also for shale gas technology, and can be referred to as a standard argument of industry against the introduction of a sector/technology-specific legal regime. The argument further reveals an interesting link between legislation and its effect on the public debate surrounding a technology. One could have expected that the public would be reassured that the introduction of a specific regulatory regime on a technology would guarantee a safer use of the latter,[539] however now the public seems to be reassured that as it is 'only' subject to existing legislation, the potential adverse effects and risks cannot be 'that bad'. As such, this line of argument is a 'dangerous' one to use, as it can mislead the public on the risks associated with a technology. As can be concluded from the comparative approach of the regulation of nanotechnology, CCS and shale gas, the regulatory 'form' or 'design' of a technology cannot be used as an indicator of its negative effects/risks associated with the technology. The choice of design has other reasons such as competence-related ones, for example.

---

[534] Marchant, note 19 above, at 25f.

[535] P. Sollie, 'On uncertainty in ethics and technology' in: P. Sollie and M. Duwell (eds) *Evaluating New Technologies: Methodological Problems for the Ethical Assessment of Technology Developments* (Dordrecht: Springer, 2009), 141–58, at 147.

[536] RPA and BiPRO, note 277 above, at 8.

[537] Ibid.

[538] Ibid., 9.

[539] Van Calster, Bowman and D'Silva, note 74 above, at 228.

## 6. CONCLUSION

This chapter has analysed the regulation of the 'new' technologies, CCS and nanotechnology and established some parallels and regulatory lessons applicable to shale gas. Regarding the legislative history and overall regulatory framework of these technologies, several conclusions can be drawn. First, concerning CCS and shale gas, it can be stated that even if the underlying motivations for the introduction of both technologies were similar (global-warming mitigation and competitiveness), different policy approaches were chosen. The application of underlying regulatory principle of conservatism, explicitly named within the context of CCS, but also applied in the context of shale gas, led to different outcomes. The Directive is the outcome of the more conservative approach; the Recommendation on shale gas, because of its form, is more liberal. A strict application of the conservatism principle in the context of shale gas would have most likely also resulted in the conclusion that a Directive was needed for the regulation of the technology, as the point of departure is similar. Some aspects are addressed by existing legislation; others still need to be regulated.[540] However, as established in Chapter 2, the horizontal division of competences makes this very difficult, if not impossible, to establish in practice. This is seemingly also the reason why the Commission did not refer explicitly to the regulatory principle of conservatism in the context of shale gas.

The regulatory history and overall framework associated with nanotechnology has established the lesson that the application of a principle-based regulation (in the form of a general code of conduct) to new technologies failed. As such, the Recommendation on shale gas, even if formally outlining minimum principles, goes beyond that. The case study of nanotechnology also established a precedent on how a Recommendation can have *de facto* and *de jure* effect. The Recommendation on the definition of nanomaterials served as the basis for the inclusion of a definition into binding acts at European level. Integrating aspects of the minimum principles into European Directives and Recommendations is certainly wishful thinking for shale gas. To date, the inclusion of 'shale gas hooks', which these arguably are, has failed as illustrated by the amendment of the EIA Directive.

The policy context of CCS and nanotechnology further shows that both technologies were embedded in a wider research context, and in the case

540 See Recitals 7 and 8 of the Recommendation itself.

of nanotechnology, also integrated into other policy areas. For shale gas, although there have indeed been multiple scientific studies on the legislative and policy frameworks, the 'soft' science, 'hard' science research projects at the European level currently do not exist. These are however needed in order to clarify existing uncertainties or risks (depending on the classification), as not all experiences can be perfectly transferred from the US to the European context.

Finally, an important difference is the general function of law regarding the three technologies. For CCS, law-makers started from the positive presumption of 'making the activity work'. Law has been an enabler of the technology; its intention was to create (financial) incentives to promote the swift introduction of CCS.[541] In retrospect one is led to conclude that the CCS Directive as it was drafted did not implement this approach. The starting point is similar for nanotechnology; regulation at an EU level was not the primary objective of these initiatives. The aim was instead to create a favourable innovative environment for the technology. Legislation was deemed to be 'possible' but not 'necessary'. For shale gas, the opposite scenario was the case. The first appearance of shale gas at European level was in relation to addressing environmental concerns rather than to promoting the activity.[542] Even if some Member States tried to lobby against European intervention, it was commonly agreed that some form of European involvement was needed. Through the form chosen for the regulatory intervention, only limited legal certainty is created for industry, policy-makers and investors. Law might have a neutral role,[543] but is certainly not an enabler of the shale gas technology. As established above, 'hard' science on shale gas in the EU is still lacking.

No correlation can be established regarding the 'newness/bindingness/principle approach' used for a technology's risk dimension. As Table 3.5 illustrates, all combinations are possible. Moreover, no correlation can be established between the newness of a technology and the application of the precaution and prevention principles for risk management. The review of the use of the principles for the three technologies further confirms that it is difficult to draw a clear line between them. This is the

---

[541] See also Woerdman and Couwenberg, note 477 above, at 97f.

[542] See for J. Potočnik, 'Transmission Note on the EU environmental legal framework applicable to shale gas', 26.01.2012, available at http://ec.europa.eu/environment/integration/energy/pdf/legal_assessment.pdf; last accessed 26.06.2015.

[543] Even though there are voices which say that technology legislation can never be neutral: C. Reed, 'Taking Sides on Technology Neutrality' 4 *SCRIPT-ed* 3 (2007), 263-284, at 279. See also for further references.

case for shale gas specifically; there, even the Commission seems to confuse the link.

Additionally, in the case of shale gas, public awareness and opposition are, when compared to the public debate surrounding the other two technologies, more significant. This is the case even if the level of uncertainty and risk involved in the technology might arguably be lower than for CCS certainly lower than for nanotechnology. Premature, negative public opinion can hinder the development potential of shale gas; despite the other two technologies having managed well on this particular issue, public opinion certainly poses a risk to shale gas. Given its current state, public opinion may well stop its development in its infancy. In nanotechnology, rising concerns were met by road shows and early information campaigns. Such measures may no longer be possible for shale gas. There, risk communication is a key issue and is what has gone wrong from the outset, at least from an operator's standpoint. Now, European policy-makers could also be about to mishandle the risk management if lessons from other technologies such as CCS and nanotechnology are not taken into account; a shift from precaution to prevention will not happen in practice.

The comparison of shale gas and other cross-cutting environmental and energy technologies shows that the aspect of 'newness' in technologies seems to become increasingly important regulatory practice. Simultaneously, however, the regulatory approach taken towards a technology does not depend on the 'newness' of a technology but on the 'newness' of the underlying science. Reverting to this finding, one can consequently also conclude that the regulatory form of a technology cannot be used as an indicator of its negative effect, risks or uncertainties. The regulatory design of a technology is ultimately based on other factors such as the division of competences. Hence, from today's perspective, a CCS measure against the background of the horizontal division of competences post-Lisbon is rendered more difficult if not impossible to establish.

The comparison further reveals that there is no coherent correlation between the elements of 'newness/bindingness/regulatory principle/public debate/business case', as the following table illustrates. Interesting to note is that – contrary to what one might have reasonably expected – public awareness and public opposition to date do not have an influence on the bindingness of a measure. Table 3.5 below illustrates and summarizes these findings.

*Table 3.5 Correlations newness/bindingness/risk management/public debate and business case*

| | CCS | Nanotechnology | Shale gas |
|---|---|---|---|
| **Degree of newness** | Adoption of an existing technology | New | Adoption of an existing technology |
| **Bindingness of measure** | Binding (Directive) | Mixture: non-binding (Recommendation); over time binding (incremental, amendments to existing legislation; Directives and Regulations) | Non-binding (Recommendation) |
| **Underlying principle** | Prevention/ Conservatism | Precaution | Call for move from precaution to prevention |
| **Public debate** | Low awareness, sporadic public opposition | Firstly cooperative approach, then opposition | High awareness and public opposition |
| **Business case** | High capital investment needed and need for financial incentives | Low capital investment and high profitability (no financial incentives needed) | High capital investment, but high profitability (no financial incentives needed) |

Comparing the regulatory design of shale gas and CCS, one can conclude, regarding public perception at least, that the form does not matter as such, but rather the overall approach to the topic, thus the use of the concept of risk, uncertainties and hazardous activities. As for CCS, the stakeholder consultation concludes that:

> the overall approach of the Directive has been to treat CCS as a hazardous activity, which has led the public to approach the topic with more caution than necessary. Some industry stakeholders (especially Oil and Gas related) mentioned that the knowledge of and experience with all elements of CCS is not well reflected in the Directive, which treats CCS as something entirely new.[544]

As such, one can conclude that the 'new' in technology does not really matter for regulation; what matters is whether the underlying science is new, which is not the case in both examples.

[544] Ricardo AEA, et al., note 203, above, at 10.

Nanotechnology on the other hand can be considered an actual new technology. The lessons learned from the regulatory experience regarding nanotechnology are twofold. The first lesson is that, even with an actual new technology such as nanotechnology, regulatory design does not necessarily lead to a coherent and comprehensive new regulatory regime. On the contrary, it can result in the adoption of non-binding tools and further fall within the scope of existing legislation, subject to technology-specific 'hooks'. This regulatory approach – combining soft-law measures and hard-law hooks to nanotechnology – has positioned the EU at 'the forefront of policy and regulatory developments in this area, and … done so with a bang'.[545]

The second lesson learned is that nanotechnology is an example of the 'proper' application of the precautionary principle. It addresses scientific uncertainties 'on the go', without slowing down the commercial development of the technology. The same might be concluded for shale gas in the future, however without any practical (exploratory) experience in the Member States, it will be impossible to assess the regulatory framework in practice. *Ex ante* regulation is certainly important, especially since we have the US experience at hand. However, as we learned from the experience with CCS, technological development and the development of the regulatory framework should go hand in hand. Indeed, contrary to the regulatory experience with the CCS technology, the regulation of nanotechnology has evolved quite well. This might indeed be due to the overarching flexible regime of Recommendations in combination with a robust and practice-proven existing legislative framework.

It remains to be seen whether the EU has learned from its regulatory history with 'new' or 'modified' technologies. Considering Collingridge, through the adoption of a non-binding Recommendation on shale gas only outlining minimum principles, the Commission might indeed have found a way to do both, promoting a technology by outlining a possible path and design of a binding measure in the future. Furthermore, with the completion of a public open stakeholder consultation[546] and the creation of a 'European Science and Technology Network on Unconventional

[545] Van Calster, Bowman and D'Silva, note 74 above, at 227. As well as Van Calster, Bowman and D'Silva, note 28 above, at 106.

[546] Analysis and presentation of the results of the public consultation 'Unconventional fossil fuels (e.g., shale gas) in Europe', Final Report, European Commission DG Environment, 03.10.2013, available at http://ec.europa.eu/environment/integration/energy/pdf/Shale%20gas%20consultation_report.pdf; last accessed 22.10.2014.

Hydrocarbon Extraction',[547] as well as the reporting obligations (or better invitations) on the application of the Commission Recommendation,[548] and the creation of a separate BREF Reference Document for shale gas,[549] the Commission indicates a move up the ladder from simply applying a precautionary approach to legislation to a preventive approach of risk management, a shift which is certainly needed and welcome.

---

[547] Suggested in the Commission Communication on the exploration and production of hydrocarbons (such as shale gas) using high volume hydraulic fracturing in the EU, Brussels, 22.1.2014, COM(2014) 23, at 1. The network was effectively launched on 8 July 2014. For more information refer to https://ec.europa.eu/jrc/en/uh-network; last accessed 22.10.2014.

[548] See Article 16(1), which invites;

> Member States having chosen to explore or exploit hydrocarbons using high-volume hydraulic fracturing to give effect to the minimum principles set out in this Recommendation by 28 July 2014, and to annually inform the Commission about the measures they put in place in response to this Recommendation for the first time by December 2014.

[549] European Commission, note 463 above; under revision until 2017.

# 4. Conclusion

The previous chapters have analysed the overall regulatory and policy framework on shale gas, departing from very specific (the case of shale gas) – over the discussion of competence-related issues – to more conceptual issues, namely the regulation of 'new' technologies. Figure 4.1 below illustrates this approach.

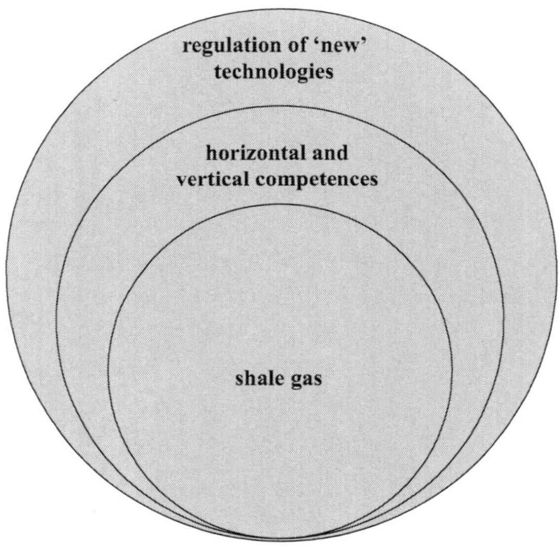


*Figure 4.1 Visualization of the analyses carried out*

Every chapter has closed with a conclusion answering chapter specific questions. Chapter 1 provided the underlying framework for the following two chapters.

Chapter 2 analysed the legal basis for environmental and energy regulation at the EU level after the Lisbon Treaty. It examined the horizontal and vertical division of competences relating to shale gas, and also stressed the importance and difficulty of the notion of the term 'significance' within this discussion. It further reviewed the impact shale gas might have on a Common European Energy Policy (CEP). It answered the following research questions and related sub-questions:

- *What is the legal relationship between Title XX on the Environment and XXI TFEU on Energy in terms of energy security of supply concerns and environmental concerns: are they really mutually compatible or does one take precedence over the other?*
    - o To what extent does the Union have competence to act regarding shale gas?
    - o How does this relation play out at a Member State level regarding each State's right to determine the conditions for exploiting its energy resources?
- *How will the issue of shale gas affect the Union's quest for a Common European Energy Policy in the absence of a Treaty provision?*
    - o Do the energy title and its opt-out mechanism in the second paragraph prevent a Common Energy Policy (on unconventional gas resources) in the future?

The chapter analysed the legal 'constitutional' issues surrounding the shale gas debate. It has illustrated the legal relationship between Title XX on the environment and Title XXI on energy and their interaction after the Treaty of Lisbon. As such it has analysed the extent to which the Union can act on shale gas, and subsequently concluded that with the non-binding Recommendation, the Commission is already acting at the brink of its competences. While the minimum principles on shale gas are an important indicator for European involvement on the issue, the fact that these are non-binding also illustrates the reluctance and possibly the anxiety of the EU Institutions to interfere in an issue which has in the past been recognized as a key aspect of Member States' sovereignty. The Commission has no exclusive competence in the environmental and energy area, and thus has to justify its action on shale gas, although not to the same extent it would have to when issuing a legislative proposal. Overall, shale gas regulation may easily pass the subsidiarity hurdle on transboundary environmental grounds, however the limitations introduced in Article 194(2) TFEU can prevent the adoption of a European (legal) measure on the issue. The reasons why the Commission acted the way it did are in essence twofold: first, the fact that Article 194 functions as a 'game changer' in the case of shale gas and prevents a European measure in its entirety, and second, the fear of 'real' subsidiarity discussions when a legislative proposal on shale gas has to go through national parliaments.

Indeed, contrary to what one would reasonably expect, legal measures under the energy competence concerning the exploration and extraction of shale gas resources are not possible; the environmental competence is the only possible legal basis for such measures. Furthermore, policy-makers in the European Institutions are expected to argue from the start that a measure is needed under the environmental competence, and not as a means to ensure energy supply; such an argument focused on ensuring security of energy supply as an overarching aim would preclude the Union from acting in this regard.

Additionally, if a shale gas measure in the form of a Directive or Regulation were to be adopted under the environmental competence, it is crucial to determine whether these shale gas measures would '*significantly* affect' or merely 'affect' the choice between energy sources and general structure of energy supply. If it would merely 'affect' the components, a measure would be subject to the ordinary legislative procedure. However, if it would '*significantly* affect' the components, it would most likely be subject to unanimity, *ergo* subject to veto rights from the Member States. This is a likely scenario as the UK and Poland are generally against European involvement. Therefore, the only way to pass a Union measure in the form of a Directive or Regulation on shale gas would be under the environment title, without a significant effect being determined, and thus on the condition that the ordinary legislative procedure is applied.

Regarding a CEP and shale gas, two scenarios are possible in respect of the existing policy. First, it is possible that proponent countries such as Poland and the UK use the existing strategies and roadmaps as an anchor point to promote and reason their engagement in shale gas activities, for example, regarding the potential of the activity to reduce greenhouse gas emissions. However, second, it is just as likely that opposing countries use the same policy documents as a basis for argumentation against shale gas on the grounds of environmental concerns and knowledge gaps. The existing policy framework is thus interpretable both ways, for and against a European voice on the issue. However, none of the strategies includes a strong recognition of shale gas as a possible game changer. Considering the strong and very divergent positions the Member States have on shale gas, it is questionable whether the current way towards a common policy can be upheld in the future. From today's point of view it seems plausible that Member States, driven by the possibility to gain energy independence, will rely on their competence to determine their energy mix and pursue their own policies regardless of pre-existing roadmaps and strategies. The new Treaty provision also does not live up to its expectations regarding the promotion of a CEP. Even if European efforts ensured the correct course in the recent past, the case of shale gas is certainly a setback to the approach of aligning European energy issues, that was taken after the gas crises. In addition, this trend could even lead to a fragmentation of European energy policy, at least on a political level, even if little has changed in legal terms. A common approach on the issue is needed, not only regarding internal energy supply security, but also in order to prevent a race to the bottom of European environmental law and policy standards.

Talus describes the EU Institutions' duty to respond to environmental concerns as a 'close to impossible task'.[1] Indeed, the new set-up of competences in the environmental and energy area is, contrary to what one would reasonably expect, complicating this issue. Instead of being a 'proper energy policy enabling clause' and facilitating the competence discussions in the energy and environmental area, the new competence seems to rather restrict the passing of energy-related regulation. The second paragraph of Article 194 TFEU did not only prove an opt-out mechanism for the Member States of European policy, but also a further hurdle to passing measures under the environmental competence, and at best only provides additional argumentation against European energy law and policy. Its reference to Article 192(2)(c) TFEU further puts weight on the significance discussion, as the unclear meaning of the wording in relation to shale gas shows. It remains to be seen how the wording evolves in the future and whether eventually the Court has to step in. It will be difficult to establish that the impacts on the energy choice will be significant.

Finally, the Chapter 3 broadened the perspective and dealt with risk and regulatory design of 'new' technologies in the EU. It compared the regulation of other 'new' technologies such as CCS and nanotechnology with shale gas in terms of the process and general direction of regulation; the risk management and the underlying principles; the business case of the technologies and lastly, the management of the public debate around the technologies. It established which lessons can be learned from or for shale gas regulation. The last two research questions were answered:

---

1 K. Talus, *EU Energy Law and Policy: A Critical Account* (Oxford: Oxford University Press, 2013), at 175.

- *What are the lessons learned from regulatory design (failures) of technologies of the past?*
- *What is the 'new' in new technologies? And how does this matter for the regulation?*

The comparison of shale gas and other cross-cutting environmental and energy technologies shows that the aspect of 'newness' in technologies seems to become increasingly important in regulatory practice; at the same time, the regulatory approach taken towards a technology does not depend on the 'newness' of a *technology* but on the 'newness' of the *underlying science*. Reversing this finding, one can consequently also conclude that the regulatory form of a technology cannot be used as an indicator of its negative effect/risks or uncertainties. The comparison further reveals that there is no correlation between the elements of newness/bindingness/regulatory principle, as Table 4.1 below illustrates.

*Table 4.1 Summary of findings*

| | **CCS** | **Nanotechnology** | **Shale gas** |
|---|---|---|---|
| **Degree of newness** | Adoption of an existing technology | New | Adoption of an existing technology |
| **Bindingness of measure** | Binding (Directive) | Mixture: non-binding (Recommendation); over time binding (incremental, amendments to existing legislation; Directives and Regulations) | Non-binding (Recommendation) |
| **Underlying principle** | Prevention/ Conservatism | Precaution | Call for move from precaution to prevention |
| **Public debate** | Low awareness, sporadic public opposition | First cooperative approach, then opposition | High awareness and public opposition |
| **Business case** | High capital investment needed and need for financial incentives | Low capital investment and high profitability (no financial incentives needed) | High capital investment, but high profitability (no financial incentives needed) |

Hence, no correlation can be established between newness of a technology and the application of the precaution and prevention principles for risk management. The review of the use of the principle for the three technologies confirmed that it is difficult to draw a clear line between them.

The case of nanotechnology is an example of the 'proper' application of the precautionary principle, as, without having slowed down the commercial development of the technology, it addresses scientific uncertainties 'on the go'. The same might be concluded for shale gas in the future,[2] however without any practical (exploratory) experience in the Member States, it will be impossible to assess the regulatory framework in practice. *Ex ante* regulation is certainly important (see findings of Chapter 2), especially since we do have the US experience at hand; although as we have learned from the experience with CCS, technological development and the development of the regulatory framework should go hand in hand.

Indeed, contrary to the regulatory experience with the CCS technology, the practical application of nanotechnology has evolved quite well. This might indeed be the case due to the overarching, flexible regime through the Recommendations, in combination with a robust and practice-proven existing legislative framework.

The underlying regulatory principle of conservatism, explicitly named in the context of CCS but also applied in the context of shale gas, led to different outcomes, as illustrated in Figure 4.2.

The Directive is the outcome of the more conservative approach; the Recommendation on shale gas, because of its form, is more liberal. A strict application of the conservatism principle in the context of shale gas would most likely also have resulted in the conclusion that a Directive was needed for the regulation of the technology, as the point of departure is similar. Some aspects are addressed by existing legislation – others still need to be regulated.[3] However, as established in Chapter 2, the horizontal division of competences makes this very difficult, if not impossible to establish in practice. This is conceivably also the reason why the Commission in the context of shale gas did not refer explicitly to the regulatory principle of conservatism. The principle has been applied without naming it as such, and it remains to be seen whether it evolves into a 'real' regulatory principle in the future. It arguably has the

[2] Just with the prevention principle as the underlying risk management principle.

[3] See Recitals 7 and 8 of the Recommendation itself.

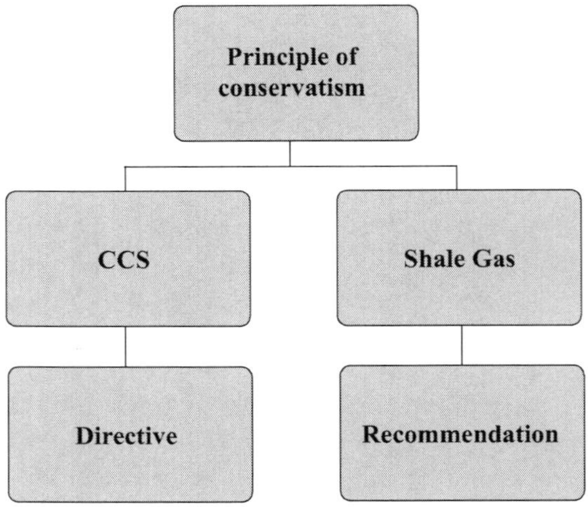


*Figure 4.2 The principle of conservatism and its application in the CCS and shale gas context*

potential as the European legislator has generally been keen on intervening regarding the regulation of 'new' technologies, certainly also because of pressure coming from the majority of the Member States. In that regard, the principle establishes 'an excuse for regulation' even in circumstances dealing with known and quantifiable risks.

The regulatory history and overall framework regarding nanotechnology have established the lesson that the application of a principle-based regulation (in the form of a general code of conduct) towards new technologies failed. As such, the Recommendation on shale gas, even if formally outlining minimum principles, goes beyond that. The case study of nanotechnology however also set a precedent on how a Recommendation can have *de facto* and *de jure* effects. The Recommendation on the definition of nanomaterials served as the basis for the inclusion of a definition into binding acts at European level. This is certainly a desirable way forward, as also for shale gas. Aspects of the minimum principles should also be included into European Directives and Regulations. However to date, the insertion of 'shale gas hooks', which these would arguably be, failed. As shown by the suggested amendment of the EIA Directive.[4]

[4] Directive 2011/92/EU of 13 December 2011 on the assessment of the effects of certain public and private projects on the environment, [2012] OJ L 26/1.

Additionally, an important difference is the general function of law within the three technologies. For CCS, law-makers started from the positive presumption of 'making the activity work'. Law has been an enabler of the technology: its intention was to create incentives to promote a fast introduction of CCS.[5] In retrospect, one is led to conclude that the CCS Directive as it was drafted, did not implement this approach in practice. Regarding nanotechnology, the starting point is similar: regulation at an EU level was not the primary focus of initiatives. The key aim was to create a favourably innovative environment for the technology. Legislation was deemed to be 'possible' but not, at first, 'necessary'. Regarding shale gas, the opposite was the case. The first appearance of shale gas at European level was in relation to addressing environmental concerns and not to promoting the activity.[6] Even if some Member States tried to lobby against European intervention, it was commonly agreed that some form of European involvement was needed. Through the form chosen for the regulatory intervention, little legal certainty is created for industry, policy-makers and investors. Law might just have an opposite and at best neutral role,[7] but is certainly not an enabler of the shale gas technology. As established above, 'hard' research on shale gas in the EU is still absent.

It remains to be seen whether the EU has learned from its regulatory history with 'new' or 'modified' technologies of the past. Considering Collingridge, through the adoption of a non-binding Recommendation on shale gas only outlining minimum principles, the Commission may indeed have found a way to both promote a technology by outlining a possible oath and designing a binding measure in the future. In addition the carrying out of a public open stakeholder consultation,[8] the creation

---

[5] See also E. Woerdman and O. Couwenberg, 'CCS in the European Emissions Trading Scheme' in: M. M. Roggenkamp and E. Woerdman (eds) *Legal Design of Carbon Capture and Storage: Developments in the Netherlands from an International and EU Perspective* (Antwerp: Intersentia, 2009), 97–125 at 97f.

[6] See for example J. Potočnik, 'Transmission note on the EU environmental legal framework applicable to shale gas', 26.01.2012, available at http://ec.europa.eu/environment/integration/energy/pdf/legal_assessment.pdf; last accessed 26.06.2015.

[7] Even though there are voices which say that technology legislation can never be neutral: C. Reed, 'Taking sides on technology neutrality' *SCRIPT-ed* 4(30 (2007), 263–84, at 279. See also for further references.

[8] Analysis and presentation of the results of the public consultation 'Unconventional fossil fuels (e.g. shale gas) in Europe' Final Report, European

of a 'European Science and Technology Network on Unconventional Hydrocarbon Extraction',[9] the reporting obligations (or better invitations) on the application of the Commission Recommendation[10] and the creation of a separate BREF Reference Document for shale gas,[11] the Commission indicates a move up the ladder from simply applying a precautionary approach to legislation to a preventive approach of risk management; a shift which is certainly needed and welcome.

These individual strains of analysis of the separate chapters are brought together in order to answer the overall research question:

- **To what extent does the regulation of 'new' technologies such as shale gas highlight the challenges of a coherent regulation of energy and environment in the European Union?**

Against the background of the analysis and conclusion of the individual chapters, the second part of this question is relatively straightforward to answer: there *is* no coherent regulation of energy and environment at EU level alone. This is mainly due to two overall reasons: first, the fact that the current approach of dealing with the *underground* is fragmented and incoherent. This is already reflected in the institutional set-up within the European Commission on how energy and underground issues are addressed. Regarding shale gas, DG Environment has the lead, DG

---

Commission DG Environment, 03.10.2013, available at http://ec.europa.eu/environment/integration/energy/pdf/Shale%20gas%20consultation_report.pdf; last accessed 26.06.2015.

[9] Suggested in the Commission Communication on the exploration and production of hydrocarbons (such as shale gas) using high-volume hydraulic fracturing in the EU, Brussels, 22.1.2014, COM(2014) 23, at 1. The network was effectively launched on 8 July 2014. For more information refer to https://ec.europa.eu/jrc/en/uh-network; last accessed 26.06.2015.

[10] See section 16(1), which invites:

> Member States having chosen to explore or exploit hydrocarbons using high-volume hydraulic fracturing to give effect to the minimum principles set out in this Recommendation by 28 July 2014 and to annually inform the Commission about the measures they put in place in response to this Recommendation, for the first time by December 2014.

[11] European Commission, BREF on Management of Tailings and Waste-Rock in Mining Activities (2009), available at http://eippcb.jrc.ec.europa.eu/reference/BREF/mmr_adopted_0109.pdf; last accessed 25.06.2015. The BAT are under revision until 2017.

ENER is only of secondary relevance. This arguably already indicates the environmentally oriented approach the Union is trying to pursue.

For shale gas, the Commission has tried to follow the familiar path of 'incremental regulation', through the inclusion of 'shale gas hooks' into existing legislation,[12] combined with a soft law approach on the issue. Depending on how this regulatory path evolves in the future, it may have to change this path and insist on its (weak) regulatory powers under the environmental title, which are likely to trigger litigation on legal basis issues. However, the Commission indeed needs to take some risks if it wants to regulate, even if such an attempt is likely to end up before the Court on a challenge on grounds of the legal basis, it is the only way forward.

The incoherent regulation of energy issues as such is also to some extent the result of the very nature of energy issues, as they are at the core of the Member States' sovereignty. This indicates the second reason for the incoherent regulation of energy and environment in the Union, namely that in both areas, the competences are shared competences with the Member States. This then leads to the next question of whether energy and environment can be regulated coherently on a multi-jurisdictional, multi-layered level? The response to this question *should* be 'yes'.

The analysis in Chapter 2 indicates that existing regulatory gaps in EU legislation regarding shale gas can be addressed; although due to competence issues, only through a non-binding Recommendation outlining minimum principles. However, as the experience with the Recommendation on the definition of nanotechnology showed, a non-binding document can also in principle achieve *de jure* effect, for example when included in existing legislation. Even if the official assessment of the transposition of the Recommendation into the existing regimes of the Member States is not completed as of yet, the regulatory initiatives in the UK (the 2015 Infrastructure Bill), as well as in Germany (the Draft Act on fracturing) suggest implementation of the Recommendation into binding national law.

In short, it can be concluded that the regulatory framework is there, under the pretence that the Recommendation is transposed into binding legislation in the Member States. Similar to the case of CCS, clarification of the regulatory framework was needed to trigger commercial interest. It is now up to citizens and politicians in the individual Member States to decide whether shale gas development will be carried out. This will also

---

12 Which have, however, failed so far, see Chapter 3.5.1.4.

test whether the *ex ante* legal framework is elaborate enough. As it has been summarized correctly in the case of nanotechnology: 'there is no a priori answer. ... The answer is what the evolving conception of Good Life will reveal through argument and trial and error.'[13] Indeed, in the case that the regulatory framework does not prove to be sufficient, 'regulation on the go', as carried out regarding nanotechnology, is the way forward. The framework on shale gas as it stands at a European level is able to be revised in parallel with gaining knowledge of the activity. The million dollar question of how safe is safe enough,[14] will not be answered without practical experience of the issue.

The first part of the overall research question is partly the answer to the sub-questions of the third chapter. It is in essence a matter of how risk is framed.[15] It has been suggested to use the worst-case scenario as a point of reference for action.[16] Although this has little to do with reality, it is often done to put forward interest-driven declarations and ideology.[17] Accordingly, 'effective regulation cannot afford to be based on imagined futures that are not connected to scientific, social and economic reality'.[18] Indeed, the discussion on shale gas in the EU has focused on the quality of science, scientific uncertainties regarding the technology, as well as worst practices in the US. In the shale gas regulatory debate, '[s]cience [indeed] simply becomes recognized as a necessary, rather than

---

13 J. C. Pitt, 'Anticipating the unknown: the ethics of nanotechnology' in: F. Jotterand (ed.) *Emerging Conceptual, Ethical and Policy Issues in Bionanotechnology* (Dordrecht: Springer, 2008), 103–16, at 114.

14 T. K. Epprecht, 'Producing safety or managing risks? How regulatory paradigms affect insurability' in: G. A. Hodge, D. M. Bowman and A. D. Maynard (eds) *International Handbook on Regulating Nanotechnologies* (Cheltenham: Edward Elgar Publishing, 2010), 103–16, at 165.

15 K. Ludlow and P. Binks, 'Regulating risk: the bigger picture' in: ibid., 144–62, at 144.

16 H. Jonas, 'The heuristics of fear' in: M. Kranzberg (ed.) *Ethics in an Age of Pervasive Technology* (Boulder, Colorado: Westview Press, 1980) 213–21, see also C. R. Sunstein, *Laws of Fear – Beyond the Precautionary Principle*, (Cambridge: Cambridge University Press, 2005), at 6.

17 A. Grunwald, 'Nanoparticles: risk management and the precautionary principle' in: Jotterand, note 13 above, 85–102, at 92. This is also the case for shale gas.

18 A. D. Maynard, D. M. Bowman and G. A. Hodge, 'Conclusions: triggers, gaps, risks and trust' in: Hodge, Bowman and Maynard, note 14 above, 573–87, at 573.

sufficient, element in effective social decision-making on risk'.[19] However, in the end, the claimed 'uncertainties' are rather known unknowns and should be subject to the prevention principle.

Furthermore, in today's globalized and mediatized world, the element of 'newness' is especially important and interesting for marketing purposes (of a product or indeed a technology).[20] As such, nanotechnology has been advertised with promises regarding its novelty and benefits.[21] However, the effect can go both ways: whereas in marketing handbooks 'newness' and 'innovation' of a product are promoted as selling points, if the element of 'newness' is associated with 'risks' or 'uncertainties', the public might rather fear the innovation and reject its application. This is what happened in the case of shale gas: instead of associating the technology as a 'new' solution to ease increasing energy demand, people fear associated risks and uncertainties. Operators would have been better off promoting the technology as 'old' at the beginning, instead of now backpedalling.[22]

Indeed, this volume studied not only the regulation of a technology, but also the technology of regulation.[23] However, 'no theory of law and

---

[19] A. Stirling, 'Science, precaution, and the politics of technological risk', *Annals of the New York Academy of Sciences* 1128(1) (2008), 95–110, at 98.

[20] See also L. B. Moses, 'Why have a theory of law and technological change?' *Minnesota Journal of Law, Science & Technology* 8(2) (2007), 589–606, at 598.

[21] J. Stilgoe, 'Governing nanotechnology in the EU: responsibility beyond risk', presented at the ACELG Workshop on Regulating risks in the European Union – the co-production of expert and executive power, Amsterdam, 21–22.05.2015), at 3. He puts forward that: 'Nano, like other technologies, has been oversold with a set of promised novelties, just as nuclear was "too cheap to meter" and agricultural biotechnology would "feed the world".' See also S. Rayner, 'The novelty trap: why does institutional learning about new technologies seem so difficult?' *Industry and Higher Education* 18(6) (2004), 349–55, as well as G. Van Calster and D. Bowman, 'Regulatory design for new technologies: Spaghetti Junction or Bauhaus Principles for regulating innovative cosmetic products XXV *Notizie di POLITEIA* 94 (2009), 75–93, at 75. As well as D. A. Kysar, 'It might have been: risk, precaution and opportunity costs' *Journal of Land Use & Environmental Law* 22(1) (2006), 1–57, at 2.

[22] The same was true for nanotechnology. In the end, nanotechnology was held to be 'just chemicals', in the same way that shale gas is sometimes referred to as being 'just drilling'. See for nanotechnology: Stilgoe, note 21 above, at 3, as well as Grunwald, note 17 above, at 96.

[23] See also R. Brownsword, 'The age of regulatory governance and nanotechnologies' in: Hodge, Bowman and Maynard, note 14 above, 60–82, at 77.

technology will answer all legal questions for all technologies'.[24] Along these lines, the volume attempted to deal with the different topics addressed in the individual chapters in a concise manner, initially there are several issues which could have have been addressed in more detail. These issues would have gone beyond the initial focus of the research. A comparison of not only the shale gas value chain with the US's regulatory regime, but also an extension of the comparative analysis of the status and use of the precautionary and prevention principle in the US, would have been an interesting exercise to undertake, also in light of the TTIP (Transatlantic Trade and Investment Partnership) agreement. The comparison would have focused on the question whether these principles play arose in the US's regulatory debate on shale gas; and if so, how this differs from the European approach. Furthermore, a comparison of the regulation of CCS and nanotechnology in both jurisdictions (US and EU), as well as in the Member States (as has been done for shale gas) would have been interesting in order to establish how the non-binding (and also binding) European elements are transposed at national level, and whether these create a coherent approach. In addition, a more detailed discussion of the business case for shale gas and the underlying economic considerations and questions on liability, insurance and the idea of creating shale gas bonds, as discussed in the US, are of practical evidence.

These issues will hopefully be included in a research agenda for the regulation of new technologies, such as shale gas. It is hoped that this volume can serve as a starting point for future research in this exciting and emerging field of law.

[24] Moses, note 20 above, at 594.

# Index